10th International Conference on Environmental Science and Technology (ICEST 2019)

AA007308

IOP Conference Series: Earth and Environmental Science
Volume 356

Xiamen, China
7 - 9 June 2019

ISBN: 978-1-7138-0152-8
ISSN: 1755-1307

Printed from e-media with permission by:

Curran Associates, Inc.
57 Morehouse Lane
Red Hook, NY 12571

Some format issues inherent in the e-media version may also appear in this print version.

This work is licensed under a Creative Commons Attribution 3.0 International Licence.
Licence details: http://creativecommons.org/licenses/by/3.0/.

No changes have been made to the content of these proceedings. There may be changes to pagination and minor adjustments for aesthetics.

Printed with permission by Curran Associates, Inc. (2020)

For permission requests, please contact the Institute of Physics at the address below.

Institute of Physics
Dirac House, Temple Back
Bristol BS1 6BE UK

Phone: 44 1 17 929 7481
Fax: 44 1 17 920 0979

techtracking@iop.org

Additional copies of this publication are available from:

Curran Associates, Inc.
57 Morehouse Lane
Red Hook, NY 12571 USA
Phone: 845-758-0400
Fax: 845-758-2633
Email: curran@proceedings.com
Web: www.proceedings.com

TABLE OF CONTENTS

ENERGY AND SUSTAINABLE DEVELOPMENT

ENVIRONMENTAL SCIENCE AND ENGINEERING

WATER TREATMENT AND WATER RESOURCES

IOP Conf. Series: Earth and Environmental Science **356** (2019) 011001 doi:10.1088/1755-1315/356/1/011001

PREFACE

It is our great pleasure to welcome you to 2019 10th International Conference on Environmental Science and Technology (ICEST 2019) which was successfully held in Xiamen Ocean Vocational College, Xiamen, China during June 7-9, 2019 .ICEST 2019 is co-organized by Xiamen Ocean Vocational College and Xiamen University of Technology, assisted by International Society for Environmental Information Sciences (ISEIS), Hong Kong Chemical, Biological & Environmental Engineering Society (HKCBEES), Environment and Agriculture Society (EAS), Fujian Smart City Association.

ICEST 2019 is dedicated to issues related to environmental science and technology. The major goal and feature of the conference is to bring academic scientists, engineers, industry researchers together to exchange and share their experiences and research results, and discuss the practical challenges encountered and the solutions adopted. Prof. Changping Chen from Xiamen Ocean Vocational College, China gave the opening remark; Prof. Caterina Valeo from University of Victoria, Canada gives her keynote speech" Scaling in Sustainable Urban Design"; Prof. Yongping Li from Xiamen University of Technology, China gives her keynote speech "Water-energy-food nexus: Integrated simulation-optimization approach"; Prof. Lixiao Zhang form School of Environment, Beijing Normal University gives his keynote speech" Food-Energy-Water Nexus for Urban Sustainability: Conceptual Framework and Real Challenges"; Prof. Guangwei Huang from Sophia University, Tokyo, Japan gives his keynote speech" Shallow Lake Restoration: Case Studies in Japan"; Prof. R. J. (Dick) Haynes from The University of Queensland, St Lucia,Queensland gives his keynote sppech" Sustainable revegetation of bauxite processing residue"; Prof. Gordon Huang from Faculty of Engineering and Applied Science,University of Regina, Canada gives his keynote speech "Management of Watershed Environmental Risks" .

Many researchers, engineers, academicians as well as industrial professionals from all over the world have presented their research results and development activities. There were five sessions: Modeling of Environmental Systems, Environmental Pollution Control, Environmental Management and Planning, Workshop on Environment, Water and Energy, Forum on South-South Sustainable Development. It will be a golden opportunity for the students, researchers and engineers to interact with the experts and specialists to get their advice or consultation on technical matters, sales and marketing strategies.

This conference proceeding presents a selection from papers submitted to the conference from universities, research institutes and industries. All of the papers were subjected to peer-review by conference committee members and international

Content from this work may be used under the terms of the Creative Commons Attribution 3.0 licence. Any further distribution of this work must maintain attribution to the author(s) and the title of the work, journal citation and DOI.
Published under licence by IOP Publishing Ltd

IOP Conf. Series: Earth and Environmental Science **356** (2019) 011001 doi:10.1088/1755-1315/356/1/011001

reviewers. The papers selected depended on their quality and their relevancy to the conference. The volume tends to present to the readers the recent advances in the field of environmental science and technology and various related areas, such as global environmental change and ecosystems management, climate and climatic changes, global warming, ozone layer depletion, carbon capture and storage, biofuels,integrated ecosystems management, satellite applications in the environment, environmental restoration and ecological engineering etc...

We would like to thank all the authors who have contributed to this volume and also to the organizing committees, reviewers, speakers, chairpersons, sponsors and all the conference participants for their support to ICEST 2019.

Prof. Gordon Huang
Faculty of Engineering and Applied Science, University of Regina, Canada
June 17,2019

IOP Conf. Series: Earth and Environmental Science **356** (2019) 011001 doi:10.1088/1755-1315/356/1/011001

Statement of Peer Review

All papers published in this volume of *IOP Conference Series: Earth and Environmental Science (EES)* have been peer reviewed through processes administered by the proceedings Editors. Reviews were conducted by expert referees to the professional and scientific standards expected of a proceedings journal published by IOP Publishing.

Organizing Committee

Conference Chair

Changping Chen, Xiamen Ocean Vocational College, China
Yongping Li, Beijing Normal University, China

Technical Program Committee

Chairs

Gordon Huang, International Society for Environmental Information Sciences
Dick Haynes, The University of Queensland, St Lucia, Queensland, Australia

Members

Prof. Selda Tekin Özan, Süleyman Demirel University, Turkey
Prof. Gordon Huang, University of Regina,Canada
Jen-Jeng Chen, Tajen University, Taiwan
Radu Godina, University of Beira Interior, Portugal
J. Helan Chandra, Department of Biotechnology, Jeppiaar Engineering College, Chennai, India
Dina M. R. Mateus, Instituto Politécnico de Tomar Estrada da Serra, Tomar, Portugal
Assoc. Prof. Gu Qiyuan, General Research Institute for Nonferrous Metals Engineering, China
Dr. Manish Vashishtha, Malaviya National Institute of Technology, Jaipur-302017 (INDIA)
AHMAD ZUHAIRI ABDULLAH, University Sains Malaysia
Faridah Othman, University of Malaya, MALAYSIA
Henrique J. O. Pinho, Instituto Politécnico de Tomar, Portugal
Kevin Fong-Rey Liu, Ming Chi University of Technology,Taiwan
Mule Mahadeo Baliram, B. A. M. University,India
Yee-Chung.Jin, University of Regina, Canada.
Sahadev Sharma, University of Malaya, Malaysia

Organizing Committee

Chair

Xuegeng Sun, Xiamen Ocean Vocational College, China
Haiyan Fu, Xiamen University of Technology, China

Member

Jing Chen, Xiamen Ocean Vocational College, China
Huiyan Cheng, Xiamen University of Technology, China
Lizhen Huang, Xiamen Ocean Vocational College, China
Yanfeng Li, Beijing Normal University, China
Jing Liu, Xiamen University of Technology, China
Guoxin Su, Xiamen Ocean Vocational College, China
Xiaoqing Chen, Xiamen University of Technology, China
Aili Yang, Xiamen University of Technology, China
Qiqi Yu, Xiamen Ocean Vocational College，China
Pan Zhang, Xiamen Ocean Vocational College, China
Shan Zhao, Shandong University, China

Content from this work may be used under the terms of the Creative Commons Attribution 3.0 licence. Any further distribution of this work must maintain attribution to the author(s) and the title of the work, journal citation and DOI.
Published under licence by IOP Publishing Ltd

Rubing Zheng, North China Electric Power University, China
Xiaocui Zhou, Xiamen Ocean Vocational College, China
Yiyong Zhou, Xiamen Ocean Vocational College, China

Peer review statement

All papers published in this volume of *IOP Conference Series: Earth and Environmental Science* have been peer reviewed through processes administered by the proceedings Editors. Reviews were conducted by expert referees to the professional and scientific standards expected of a proceedings journal published by IOP Publishing.

Content from this work may be used under the terms of the Creative Commons Attribution 3.0 licence. Any further distribution of this work must maintain attribution to the author(s) and the title of the work, journal citation and DOI.
Published under licence by IOP Publishing Ltd

IOP Conf. Series: Earth and Environmental Science **356** (2019) 012001 doi:10.1088/1755-1315/356/1/012001

Validation of wind turbine wake environment modelling

Q Chen[1] and X Zhang[2]

1 Wind Engineering Research Center, Xiamen University of Technology, Xiamen, Fujian, 361024, China
2 School of Environmental Science and Engineering, Xiamen University of Technology, Xiamen, Fujian, 361024, China
Email: 316994464@qq.com

Abstract. The widely used wake environment models in engineering are validated according to wind tunnel experiment in this paper, namely the Jensen wake model, Frandsen wake model, the Larsen wake model and simplified vortex wake model. From the validation results, it can be found that the wake expansion are overestimated, namely the expansion efficient still needs to be further investigated. In Jensen and Frandsen models, the assumption of uniform distribution of wake speeds along radius differs from the experimental data, with low precision. The wake speed with an approximate Gaussian distribution adopted by Larsen model improves the accuracy. But for all these wake models, due to the overestimation of wake expansion, the wake profiles are definitely calculated with error, thereby lowering the precision of power prediction of wind farms. It is suggested that these models needs to be improved based on more measurement/experimental data.

1. Introduction

The arrangement of wind turbines in a wind farm (namely, microscopic site selection) is closely related to the power generation efficiency of the wind farm. The reason is that when the incoming wind flow passes through the wind turbine upstream, the energy of the wind turbine downstream is weakened, which affects the output of the wind turbine in the rear row, which is called the wake effect. According to the research, the influence of wake on the output of wind farm can be as high as 10-20%[1].

In order to estimate the influence of wake, some scholars put forward wake environment modelling, but the calculation results of these wake models are quite different, so it is urgent to verify and analyze the accuracy of wind farm wake effect prediction, providing references for practical engineering application. In this paper, four wake models commonly used in engineering are compared and evaluated through wind tunnel experimental data.

2 Wake models

At present, the wake models widely used in engineering include Jensen wake model[2, 3], Frandsen wake model[4], Larsen wake model[5] and simplified vortex wake model[6].

(1) Jensen wake model[2, 3]

Jensen's model was proposed by Røs of the national laboratory of Denmark in 1983, and was adopted by the wind resource assessment software WAsP. This model assumes that the expansion of wake presents a linear change, and the expression of wake influence radius R and wake velocity u_w is shown in equation (1) and (2) :

$$R = R_0 + k_1 x \tag{1}$$

Content from this work may be used under the terms of the Creative Commons Attribution 3.0 licence. Any further distribution of this work must maintain attribution to the author(s) and the title of the work, journal citation and DOI.
Published under licence by IOP Publishing Ltd

$$u_w = u_\infty \left(1 - \left(1 - (1 - C_T)^{\frac{1}{2}} \right) \left(\frac{R_0}{R_0 + k_1 x} \right)^2 \right) \tag{2}$$

In the above formula, R_0 is the radius of the wind turbine, expansion coefficient k_1=0.075, x is the distance from the downstream of the wind turbine, u_∞ is the free flow velocity, and C_T is the thrust coefficient of the wind turbine.

(2) Frandsen model[4]

Frandsen wake model is proposed based on momentum theory, and wake velocity U_w is expressed as

$$\frac{U_w}{U_\infty} = \frac{1}{2} + \frac{1}{2}\left(1 - 2\frac{A_0}{A} C_T \right)^{\frac{1}{2}} \tag{3}$$

The model defines wake diameter D(x) as

$$D(x) = \left(\beta^{k_2/2} + \alpha s \right)^{1/k_2} D_0 \tag{4}$$

D_0 is the diameter of wind turbine, $A_0 = \pi D_0^2/4$, $A = \pi D^2/4$, $s = x/D_0$, k_2=3, $\alpha = 0.5$, $\beta = 0.5\left(1+\sqrt{1-C_T}\right)\left(1-C_T\right)^{-1/2}$.

(3) Larsen wake model[5]

Larsen wake model is a semi-analytical model, which is derived based on the turbulence boundary layer theory of Prandtl and includes first-order and second-order models[5,6].

(4) Simplified vortex wake model[7]

The vortex model is based on the research on the wake vortex structure of wind turbines. However, considering that the calculation of wake evolution process is complex, it is difficult to be directly applied to engineering practice. Therefore, simplified vortex model is often adopted, as follows:

$$U_w = U_\infty \left(1 - a - af\left(1 + f^2\right)^{-1/2} \right) \tag{5}$$

$$R = R_0 \left(1 - Ef\left(1 + f^2\right)^{-1/2} \right)^{-1/2} \tag{6}$$

where $f = x / R_0$,a is axial induction factor, $E = a/(1-a)$.

3. Experimental setup

The experimental measurement of wind turbine wake was carried out in an open wind tunnel of Delft University of Technology in the Netherlands. In the experiment, a two-blade horizontal axis wind turbine model (with a diameter of 60cm) was adopted, and Stereoscopic Particle Image Velocimetry (SPIV) was used to collect the images of wake flow field[8,9]. The experimental data were collected randomly when the wind turbine blade rotates based on the SPIV system and then averaged. The experimental tip speed ratios were 4.8 and 6.0 respectively, and the wake flow data of 2.7 times of wind turbine diameter range along the downstream direction of the wind turbine were collected.

4. Analysis and validation

The wake models are analyzed based on wake boundary and wake velocity profile. The prediction of the radial influence range of wake is an important prerequisite for wind turbine layout in wind farms, and the wake velocity profile is also an important index for evaluating the attenuation and recovery of the wake flow. The combination of the two aspects can provide a direct basis for the micro-site selection of wind turbines and an important content of the wake calculation model.

In table 1, the similarities and differences of each wake model referring to wake radial boundary and wake velocity profile are summarized. It can be seen from the table that Jensen model only assumes that the wake boundary expands linearly along the downstream direction, while Jensen model and Frandsen model both assume that the wake velocity does not change along the radius.

IOP Conf. Series: Earth and Environmental Science **356** (2019) 012001 doi:10.1088/1755-1315/356/1/012001

Table 1. Comparisons between wind turbine wake analytical models

Wake model	Wake radial boundary	Wake velocity profile
Jensen model	Linear expansion along the downstream	Distributed uniformly along the radial direction
Frandsen model	Nonlinear expansion along the downstream with initial expansion at wind turbine location	Distributed uniformly along the radial direction
Larsen model	Nonlinear expansion along the downstream with initial expansion at wind turbine location	Distributed approximately Gaussian along the radial direction
Simplified vortex wake model	Nonlinear expansion along the downstream	Only the velocity distribution on the wake center axis is considered

Based on the experimental data, the calculation accuracy of wake models are compared and verified.

4.1 Wake expansion

Figure 1 compares the predicted wake radial boundary distribution with the experimental values by Jensen model, Frandsen model, Larsen model and simplified vortex model with the tip-speed-ratio TSR of 4.8 and 6.0. In figure 1, the x-axis represents the distance from the position of the wind turbine to the downstream (in dimensionless form x/D), and the y-axis represents the radial distance from the hub of the wind turbine (in dimensionless form y/R). The wake boundary in the experiment is determined at the location when the wake speed reaches $99\%u_{\infty}$.

When TSR=4.8 (see figure 1(a)), the wind turbine wake measured in the experiment gradually expands outward downstream from y/R=1, that is, the blade tip, and reaches the maximum radius (y/R is about 1.2) at x/D=2, then roughly remaining unchanged. It can be seen from figure 1(a) that the boundary of the simplified vortex model also shows a trend that the boundary expands radially and then remains stable, but its calculated value is about 8.2% higher than the experimental value. Jensen model assumes that the wake boundary continues to expand linearly downstream from y/R=1(at x/D=0). At x/D=2.7, it expands to y/R=1.4, about 16.7% higher than the experimental value. As shown in table 1, Jensen model is based on the premise that the wake boundary expands linearly. Both Frandsen model and Larsen model set the wake boundary where the wind turbine is located to have initial expansion, and the influence range y/R is around 1.3(at x/D=0). As can be seen from figure 1, both of them are greatly deviated from the experimental value, while the continuous expansions along the downstream with linear rule also deviate from the experimental data which brings significant errors to the horizontal spacing arrangement of wind turbines in the actual wind farm.

When TSR=6.0 (see Figure 1(b)), the measured wake boundary from the experiment is more significant than that of TSR=4.8, and the wake radius y/R at the position x/D=2.5 reaches 1.3. It can be seen that the radial range of the wake flow increases with the increase of the tip-speed-ratio. The wake boundary calculated by Jensen model does not take into account the influence of the tip speed ratio. The predicted result of Jensen model is the same as that of TSR=4.8. The other three models all take the influence of the tip speed ratio into consideration. It can be obtained from figure 1(b) that at x/D=2.5, the calculation result of the simplified vortex model is about 7.7% higher than the experimental value, the calculated results of Frandsen model and Larsen model are about 23% and 29% separately higher than the experimental value.

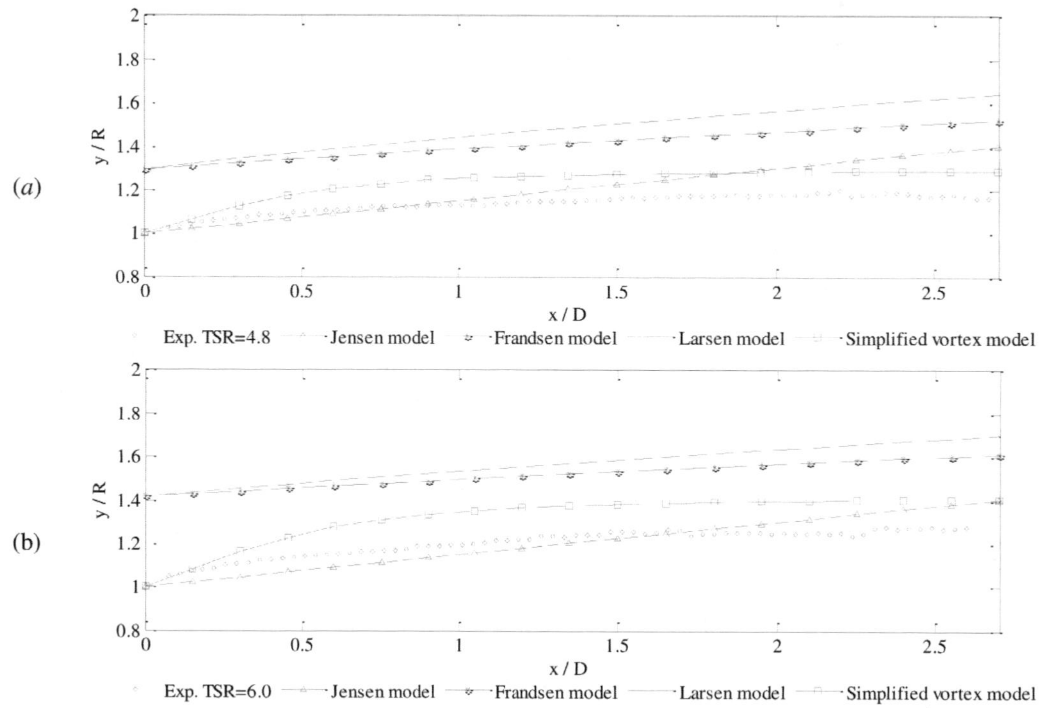

Figure 1. Validations of wake expansion (*a*) TSR=4.8 (b) TSR=6.0

4.2 wake velocity profiles

In the micro-site selection of wind farm, in addition to the prediction of radial influence range of wake, the velocity field in the downstream development process of wake is also one of the key bases for d wind turbine layout.

Figure 2 shows the radial distribution of the velocity at different downstream positions(x/D=0.5, 1, 1.5, 2, 2.5) (represented by dimensionless form u_w/u_∞). Near the wake center axis, the velocity attenuation is significant, and u_w/u_∞ is about 35%, which is consistent with the momentum theory of blade element[1]. In the range near the blade tip (0.6<y/R<1.2), with the development of the wake to the downstream (from x/D=0.5 to 2.5), the wake speed gradually recovered. However, the experimental data collected by the time-averaged technique are different from the phase-locked technique, so the data of the two are slightly different. However, it can be clearly observed that the velocity near the blade tip increases as the wake extrapolates downstream.

When TSR=4.8 (see figure 2(a)), first of all, Jensen model and Frandsen model assume that wake velocity u_w does not change in the radial direction and becomes u_∞ abruptly at the wake boundary. Both these two models overestimate the speed recovery, Jensen model results are closer to the experimental value than that of Frandsen model but still about 25% higher than the experimental value, namely underestimated the wake effect. Second, the velocity Gaussian distribution law of Larsen model (first and second order) is close to the experimental data. In particular, Larsen model presents a relatively ideal prediction result at x/D=2.5 (in the region of y/R<1.2), and Larsen's first-order model is more consistent with the experimental value than the second-order model in the region of y/R>1.2. Finally, from the calculation results of simplified vortex model on the wake center axis velocity, the predicted value is slightly higher than the experimental value about 5%~10%. According to the velocity data on the wake center axis, only simple prediction can be made for the downstream wake flow field.

When TSR=6.0 (see figure 2(b)), the predicted accuracy of the other three models is not ideal, except that the prediction of the velocity at the wake center axis by the simplified vortex model is in good agreement with the experimental value by comparing the calculated value with the experimental value of each model. First of all, at the downstream positions of x/D=0.5 and 1, although Jensen model can better reflect the wake velocity distribution, but due to the interference between the consideration of wind turbines, wind turbines with engineering in the front row layout of the longitudinal spacing will not less than twice the diameter, and with the increase of the downstream distance *x*, prediction accuracy of Jensen model is not better than TSR=4.8 condition. In addition, the calculation results of Frandsen model deviate from the experimental data to a large extent, and the prediction accuracy of Larsen model (first and second order) is lower than that of TSR=4.8 condition.

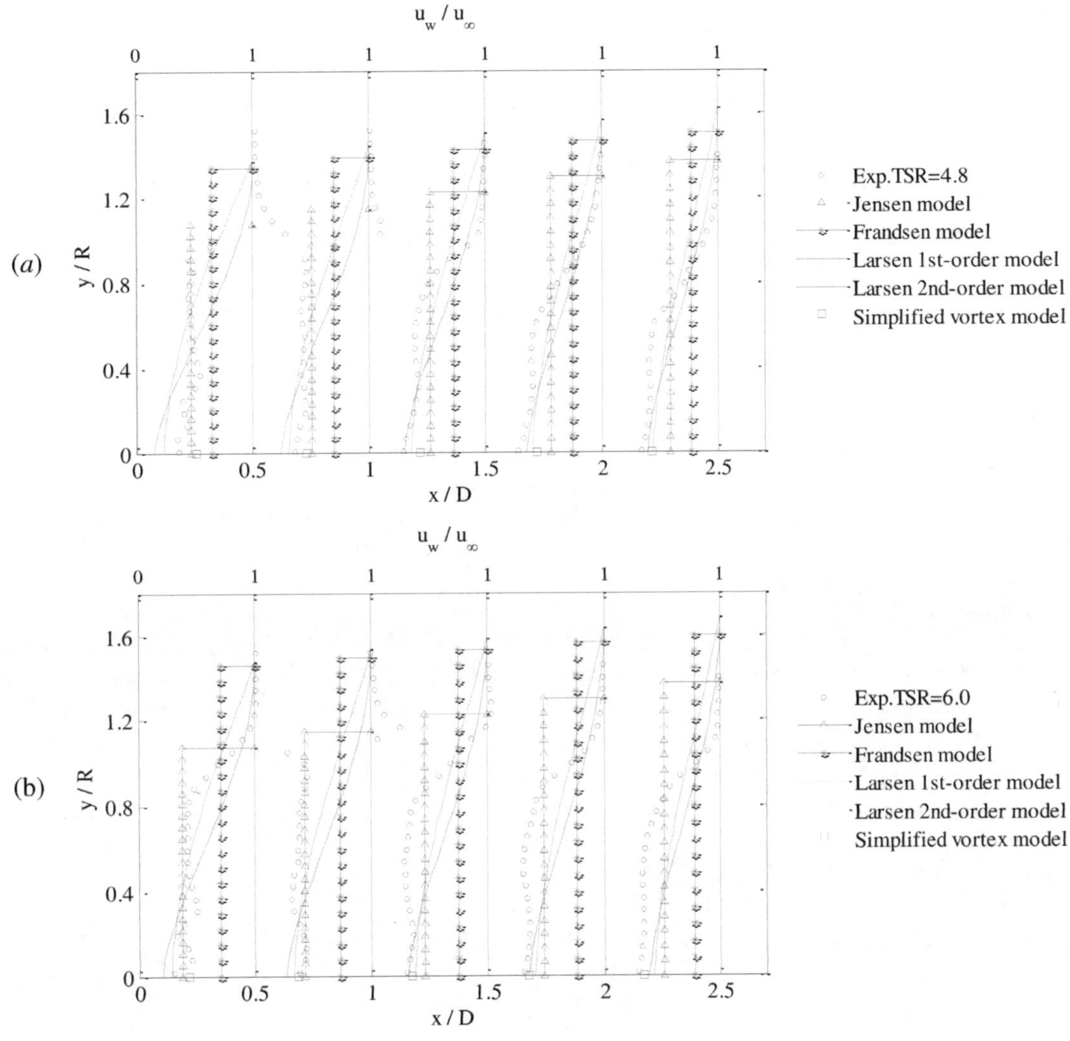

Figure 2. Validations of wake speed profiles u_w/u_∞ (*a*) TSR=4.8 (b) TSR=6.0

5 Conclusion

In this paper, the wake models commonly used in current engineering are verified and analyzed based on experimental data. The conclusions can be drawn as follows:

(1) The wake boundary expansion rate behind wind turbine varies along downstream, while Jensen model, Frandsen model and Larsen model do not make corresponding adjustment and overestimate the boundary of wake. The simplified vortex model considers the variation trend of wake expansion along the downstream, but the predicted wake expansion range is larger than the experimental data.

(2) The assumption of uniform distribution of wake velocity by Jensen model and Frandsen model is inconsistent with the actual situation. In comparison, the result of Jensen model is close to the test data within the radius of the wind turbine, while Frandsen model overestimates the speed recovery of the wake, namely underestimating the influence of wake effect. Larsen model adopts Gaussian

distribution which improves the prediction accuracy of wake velocity profile, however due to the prediction error on wake boundary, the velocity distribution presents a large deviation outside the radius of the wind turbine.

To sum up, there are still many shortcomings in the wake models applied in engineering. In order to improve the accuracy of engineering application, the calculation of wake influence boundary and wake velocity profile should be optimized. In addition, the wake experiment data used for verification also need to be further supplemented, so as to better verify and improve the wake model.

6.References

[1] Spera D A. Wind turbine technology: fundamental concepts of wind turbine engineering[M]. New York: ASME Press, 1994: 638.
[2] Jensen N O. A note on wind generator interaction[R]. Riso National Laboratory, Roskilde, 1983.
[3] Katic I, Hojstrup J, Jensen N O. A simple model for cluster efficiency[C]. Rome, Italy: 1986.
[4] Frandsen S, Barthelmie R, Pryor S, Rathmann O, Larsen S, Hojstrup J, Thogersen M. Analytical modelling of wind speed deficit on large offshore wind farms[J]. Wind Energy. 2006(9): 39-53.
[5] Larsen G C. A simple wake calculation procedure[M]. 1988.
[6] Larsen G C, Hojstrup J, Madsen H A. Wind fields in wakes[J]. 1996.
[7] Wilson R E. Wind turbine flow field model[J]. Transactions of the ASME. 1986, 108: 344-345.
[8] Qiuhua Chen, Xu Lai, SPIV analysis of correlation of tip vortex and wake flow of horizontal axis wind turbine [J]. Acta energiae solaris sinica. 2016, 2(37): 297-302.
[9] Lignarolo L E M, Ragni D, Krishnaswami C, Chen Q, Ferreira C J S, van Bussel G J W. Experimental analysis of the wake of a horizontal-axis wind-turbine model[J]. Renewable Energy. 2014, 70: 31-46.

Acknowledgement

The authors would like to acknowledge with great gratitude for the supports of the national science foundation of China(grant No: 51778551), the science and technology plan project of Xiamen(grant No: 3502Z20161016), the young teachers education science research project of Fujian(grant No: JAT170410), the Natural Science Foundation of Fujian Province(grant No. 2018J01525), and the projects of Xiamen University of Technology(grant No. XPDKQ18034 & XPDKQ18042).

IOP Conf. Series: Earth and Environmental Science **356** (2019) 012002 doi:10.1088/1755-1315/356/1/012002

The Role of Muhammadiyah in Sustainable Development Through Community Empowerment Program

M Ahmad[1] RB Soemanto[2] TK Drajat[2] W Edi[2]

1.Student of Postgraduate School, Sebelas Maret University, INDONESIA
1.Departement of Economics,University of Muhammadiyah Yogyakarta INDONESIA
2.Lecturer of Postgraduate School, Sebelas Maret University, INDONESIA
Email: macrov_jogja@yahoo.com

Abstract. Sustainable development is the development goal of every country, including Indonesia. The process of sustainable development has positioned the community as the subject with the main actor of the development is the community itself. Regarding this context, the process is performed through community empowerment approach. Muhammadiyah is the biggest non-governmental organization in Indonesia which engaged in religion, social community, economy, education, environment, and health sectors. This study aimed to discover the role of Muhammadiyah in the sustainable development of Indonesia. The method used in this study was mixed method as the combination of quantitative and qualitative approaches. The results of the research have concluded that Muhammadiyah has positive and significant roles in sustainable development of Indonesia, including from the evaluation results of the restoration program of tropical forest ecosystem in East Borneo Province and social-economic empowerment on vulnerable groups in Yogyakarta Province. The approach of Muhammadiyah through community empowerment program has been successful to improve the knowledge, change of attitude, and skills of the facilitated community. The community empowerment approach was implemented by Muhammadiyah to encourage public participation in sustainable development.

1. Introduction

At present, it has been realized that more issues faced by the world due to the climate change: decrease of biodiversity, poverty, trust crisis and others [1]. Sustainable development is the development which committed to provide contributions for the future [2]. The Sustainable Development Goals (SDGs) is the international agenda which becomes a continuation of Millennium Development Goals (MDGs), this agenda was made to answer the demand from word leadership in overcoming poverty, gap, and climate change in the form of real actions [3]. The sustainable concept itself is at least contains two dimensions, the first is the dimension of time because sustainable means related to the condition which will occur in the future, the second is interaction between economic system and environmental-natural resources system [4]. Sustainable Development Goals (SDGs) provide opportunities as well as challenges for developing countries because the implementation cost will be expensive although it could bring long-term benefits [5].

In the implementation effort of sustainable development, the initial phase has to be started through the community empowerment as the determinant of success. Community empowerment is a process in which the efforts or potentials conceived by the community are integrated to the resources owned by the government to improve the economic, social, and cultural conditions and integrating the

Content from this work may be used under the terms of the Creative Commons Attribution 3.0 licence. Any further distribution of this work must maintain attribution to the author(s) and the title of the work, journal citation and DOI.
Published under licence by IOP Publishing Ltd

community in the context of national life as well as empowering the community to be capable of provide full contributions to achieve national-level progress [6]. The orientation of community empowerment has to support the community in order to develop themselves on the basis of the existing innovations that implemented in a participatory manner in which the method of approach has to be oriented to the needs of the facilitated community and practical things in the form of individual or group services [7], [8]. Community empowerment has to be performed through three paths, namely enabling, empowering, and protecting [9].

Indonesia is one of the developing countries which still has a serious poverty issue because until present, the poverty keep approaching the economic condition of Indonesia, therefore, it needs to be relieved or at least-reduced [10]. Poverty is a complex matter, due to that cause, the intervention from the entire parties is required. Numbers of stakeholders which include the ministries/institutions, Statistical Central Bureau, academicians, experts, civil society organizations, philanthropists, and businesses have involved in various preparation processes of SDGs' implementation. On its implementation, there is an integrative principle as the fundamental idea which contains the meanings that SDGs are implemented in integrative manner and interrelated on all social, economic, and environmental dimensions [11]. Muhammadiyah is one of the biggest organizations in Indonesia which engaged in various sectors of religion, social community, economy, education, environment, or health [12]. The movement of Muhammadiyah is taking part amidst the Indonesian nation by establishing various charity businesses which truly capable of touching the lives of many people. It has been recorded in 2018 that Muhammadiyah owns elementary schools (2,250 units), middle schools (1,111 units), high schools (1,291 units), universities (171 units), hospitals and health services (2,119 units), disabled rehabilitation (82 units), and social homes (318 units) [13].

The movement of Muhammdiyah is orientating towards the benefits of people. In running the sustainable development programs, the assignment is given toward Community Empowerment Assembly (*Majelis Pemberdayaan Masyarakat* (MPM)) and Living Environment Assembly (*Majelis Lingkungan Hidup* (MLH)). In describing the procedure of empowerment activity in various communities, MPM is referring to the ecological approach of human and environmental development which mentions that social intervention has to be able to reach the entire relational levels between individual and environment. The fields of work that empowered by MPM are emerging awareness of the community regarding the rights and responsibilities as citizen, the development of basic needs, and the improvement of the income of poor and marginalized community, policy advocacy, development of crisis management center which functioned as the anticipation of psychosocial problems in the community [14].

The programs implemented by Muhammadiyah through the activities of MPM strengthen the three main pillars of sustainable development including social, economy, and environment. The concept of community empowerment conducted by Muhammadiyah becomes the key point to motivate the involvement of the community and significant economic growth. In relation to the community empowerment to reduce poverty, this effort is achieved through the outcome of the increasing rate of community's income through institutional/organizational enhancement of group economy, and the efforts of conserving the living environment. Due to that matter, this study was aimed to discover into what extent the role of Muhammadiyah in sustainable development through the community empowerment program, particularly regarding the restoration program of tropical forest ecosystem in East Borneo Province and social-economic empowerment on vulnerable groups in Yogyakarta Province.

2. Methodology

This study was designed with mixed method as the combination of qualitative and quantitative forms through descriptive statistic analysis method [15]. The data used were primary data acquired through interview, questionnaire, and observation. The data collection method was sampling through snowball technique. The respondents in this study were 266 heads of the beneficiary family of tropical forest ecosystem restoration program in East Borneo Province and the vulnerable groups, including the disabled people community and waste-pickers in Yogyakarta Special Region.

IOP Conf. Series: Earth and Environmental Science **356** (2019) 012002 doi:10.1088/1755-1315/356/1/012002

3. Result and Discussion

In principle, the empowerment activity performed by MPM is conducted to motivate the facilitated members to have knowledge and positive mental. The facilitation activities either in the forms of socialization, training, and others are aimed to improve knowledge of the facilitated members as well as enhancing the attitude and skill, therefore, the members will be motivated to conduct effective and beneficial activities for their daily lives.

The questionnaire in this study used Likert scale to measure the answers of the respondents, there were five choices of answer that can be categorized into 5 continuum scales namely very poor (1), poor (2), fair (3), good (4), and very good (5). The results of validity test have indicated that every items from Q1 to Q15 have r-table values < r-count, thus, all of them were valid. The results of reliability test have indicated that > 0.60 value of Cronbach's Alpha has been acquired, in that order, the entire question items of the questionnaire were determined as reliable. According to the results of questionnaire filling toward the entire facilitated members of MPM PP Muhammadiyah in 2015-2018 period, there were 266 household heads who analyzed through SPSS in which the obtained results are as follows (See Table 1).

Table 1. Statistical Analysis Results of the Entire Variables

Variable	N	Minimum	Maximum	Mean	Standard of Deviation
Knowledge	266	1	5	4.1	0.492
Attitude	266	2	5	3.9	0.282
Skill	266	2	5	3.7	0.340

The statistic processing results have indicated that in mean, the answer of respondents on knowledge aspect is 4.1. The attitude aspect is 3.9 in mean. The skill aspect is 3.7 in mean. On all aspects, the standard of deviation values are lower than mean, thus, it can be concluded that the data were properly distributed (See Table 1). According to the questionnaire results which have been analyzed and supported by the result data of interview toward the head of community group facilitated by MPM, thus, the perceptions regarding the implementation results of Muhammadiyah's Programs in sustainable development activities for each aspect of knowledge, attitude, and skill are as follows:

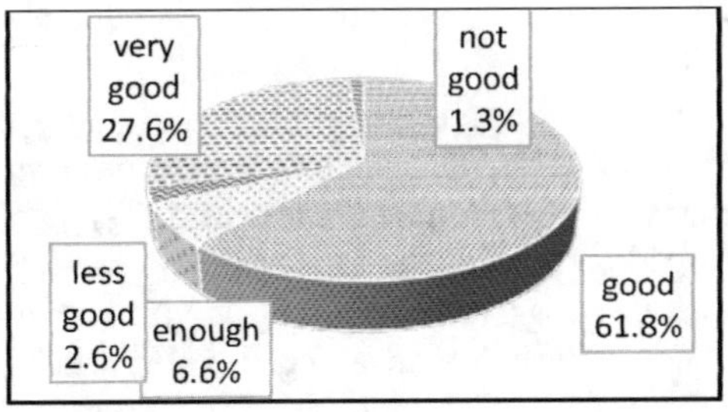

Figure 1. Improvement of Knowledge

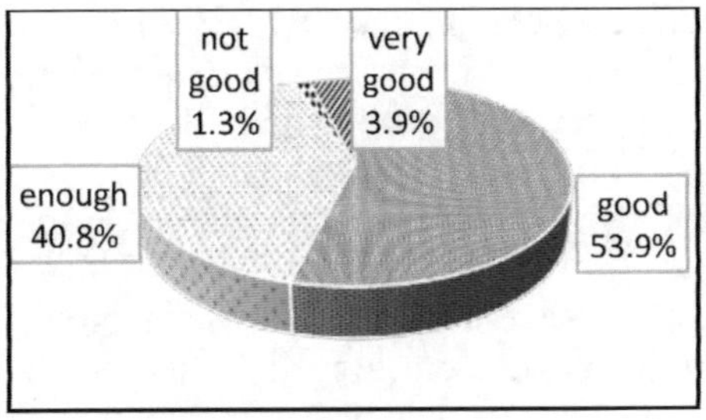

Figure 2. Change of Attitude

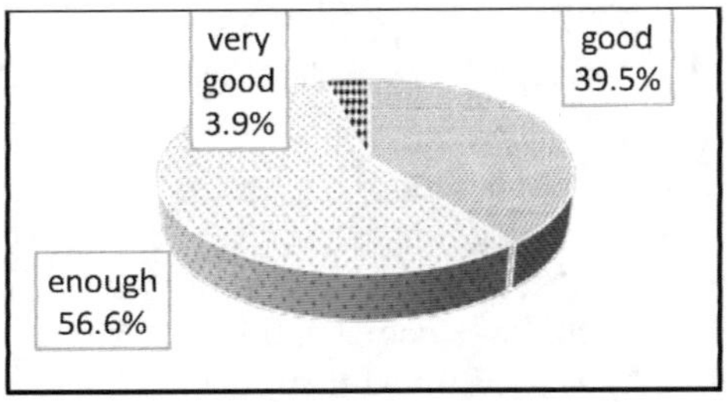

Figure 3. Improvement of Skill

The levels of knowledge, attitude, and skill of the entire members after facilitated by MPM PP Muhammadiyah were categorized as good. This condition was seen from the perception of respondents regarding the results of MPM's programs in which the majority of respondents (61.8%) have stated that there is an improvement of knowledge (see Figure 1), the majority of respondents (53.9%) have also stated that there is a positive attitude (see Figure 2), and the majority of respondents (56.6%) have mentioned that there is an improvement of skill (see Figure 3). These results have shown that the community empowerment conducted by MPM has improved knowledge, attitude, and skill of the beneficiaries of forest restoration or social-economic programs on vulnerable communities.

The main program associated with sustainable development conducted by MPM was the restoration program of tropical forest ecosystem through the development of non-timber yields to improve the income of the community and the reduction of carbon emission in East Borneo Province which has been ran in 2015-2017. This program was implemented toward the local people who live in Batu Rajang Village, Siduung Indah Village, and Long Keluh Village in Kelay and Senggah Districts that inhabit the area around forests which still in a state of poverty, therefore, poverty eradication effort is required through low-carbon economic growth, thus, 75% of the forest area of Berau Regency can be protected from deforestation, plantation, and mining.

MPM has conducted the collective action along with the community through three main activities, 1) agricultural cultivation by performing training and technical facilitation, strengthening groups as social capital, strengthening the land intensification strategy; 2) the processing of agricultural yields which initiated by the identification of potentials and value of benefits that becomes the gateway of community creative economy, the establishment of joint business group as the producing social capital, and growing the entrepreneurship mentality as the business capital, and 3) marketing process which conducted through four steps, namely assessment and establishment of cooperative as the marketing institution, training to enhance mentality and the capacity of cooperative management, administration of licensing/permit for cooperative, and building partnership with stakeholders.

This program has improved the income of household and low-carbon economic growth through the restoration of topical forest ecosystem and the development of non-timber forest products which proved by 190 households that have practiced a new system in forest management on 171 ha of area and produced non-timber products. The knowledge and skill of community in Sidung Indah, Batu Rajang, and Longkeluh Villages have improved with the evidence in which 190 people are implementing the agricultural cultivation practice around the forest area that organized in 9 farmer groups. One cooperation and 3 industrial business groups for the production forest yields have been established as the strengthening result of economic institution.

The beneficiaries in three villages of activity location own varied experience and knowledge in performing agricultural cultivation. Besides planting paddy, there is part of the community that already planted the other plantation crops, namely cocoa, rubber, pepper, agarwood, vegetables, and others.

Besides the forest restoration, the sustainable development program conducted by Muhammadiyah is community empowerment toward vulnerable groups, particularly for disabled people, waste-pickers, and street vendors in Yogyakarta Province. This form of social, economic, and environmental empowerment activities is focused on the effort of providing skill knowledge as well as personal motivation/willpower. The performed activities include training, facilitation, socialization, education, and raising involvement in organization to grow life motivation and livelihood. According to interview and observation results, it seems that there are diversity of empowerment activities conducted by Muhammadiyah on vulnerable groups.

The community empowerment through religious activities is implemented in the form of routine recitation on each group and cross-group inspirational recitation combined with mutual learning, mutual motivation, and exchange of experience as well as inspiring each other. Health facilitation, MPM uses preventive approach through socialization regarding health and curative through free medication in cooperation with PKU Muhammadiyah hospital.

The community empowerment in the sector of social economy is for the community of pedicab driver. The form of empowerment is conducted through the strengthening of pedicab driver organization through the Pedicab Driver Union of Ahmad Dahlan. In addition, the socialization regarding the regional and traffic regulations are also provided for the members to know the rights and responsibilities in traffic. Another activity that implemented is English language training, therefore, the members can understand the basic conversation to greet and offer the service and price toward the foreign tourists. MPM also provides some pedicabs to the facilitated members. This provision of pedicabs is conducted by MPM in collaboration with Lazismu.

The community empowerment on street vendor group is directed toward healthy and halal foods. The first facilitation conducted is the improvement of organizational capacity by establishing a group named as Asongan Surya Mandiri. MPM uses quality checking of food products sold by street vendors to discover whether the foods produced by them contain dangerous materials such as borax, preservative, and others. This checking/examination performed by MPM in cooperation with the laboratory of university. The facilitations implemented after the examination are in the forms of training and socialization about healthy, hygiene, and halal foods. This training activity is conducted by MPM by cooperating with Health Agency to provide facility toward the facilitated members through the provision good and healthy certification toward the foods of the members, therefore, the foods can be accepted by the schools. Another form of facilitation that conducted is the provision of merchant carts to facilitate the members in selling at schools.

The social-economic empowerment on micro business group is also implemented through the activity of organizational capacity improvement by establishing organization group. This group consists of housewives, therefore, the members organize social gathering (*arisan*) as the media to unite them in the group. The performed facilitation is the capacity improvement of low-cost production. This effort is implemented through socialization about the procedure to make proper product as well as through training regarding packaging and marketing of the product. In this training activity, MPM is cooperating with Department of Industry, Trade, and Cooperative (*Disperindagkop*). MPM has also provided the instruments that can be used by the facilitated members to perform rapid production.

Community empowerment on waste-picker community. The first facilitation that conducted is through group enhancement by establishing an organization named as Guyub Makmur. The conducted activity leads to hygienic street vendors (*angkringan sehat*), such as food-making training which has been done, including the making of sweet potato-based donut (*donat telo*), the provision of merchant carts to the group, socialization activity about the household financial management and gymnastics to establish the solidarity of members in the group.

The community empowerment has also conducted by MPM toward disabled people. The empowerment in the group of this disabled people was started through cooperative training in aimed to improve the comprehension of disabled group regarding the management of cooperative. Then, a saving and loan cooperative is established that will be ran by the members of disabled group. The facilitation activities continue to run in order to control the implementation of the cooperative. Besides establishing the saving and loan cooperative, MPM also conducts various activities to improve the welfare of disabled people, starting from the entrepreneurship training to health examination performed through cooperation with local health department.

4. Conclusion

According to the research, thus, it can be concluded that Muhammadiyah through Community Empowerment Assembly (MPM) has crucial role in community development by running tropical forest restoration program and community empowerment toward the vulnerable group such as disabled people, micro economic enterprise, street vendors, and waste-pickers as the inseparable part of sustainable development. The role of Muhammadiyah is as the facilitator, inspirer, and mediator in community empowerment program in order to improve the welfare of the local community in accordance with Sustainable Development Goals (SDGs). In addition, the results of empowerment toward poor community by Muhammadiyah is the improvement of community welfare which seen from three aspects, namely knowledge, attitude, and skill that are improving after participating in community empowerment program held by Muhammadiyah. The recommendation of the author is that the facilitated members as the implementer of empowerment program have yet to own sufficient power and independency to conduct activities in their community. Due to that matter, the role and the presence of facilitators have to be maintained until the members are fully accustomed to the empowerment process that being implemented.

5. References

[1] Broman, G. I., & Robert, K.-H. (2015). A Framework for Strategic Sustainable Development. Journal of Cleaner Production.

[2] Walkowiak, E. (1996). Sustainable development as an economic development strategy. Economic Development Review, 75–77.

[3] Financial Services Authority of the Republic of Indonesia. 2017. Sustainable Development Goals. Jakarta: Financial Services Authority

[4] Heal,G.1998 Valuing the Future: Economic Theory and Sustainability. Columbia University Press. New York.

[5] Jaiyesimi, R., 2016, The Challenge of Implementing the Sustainable Development Goals in Africa: The Way Forward, African Journal of Reproductive Health, Pp. 13-18.

[6] Einsidel, Luz, A, 1968. Succes and Failure of Some Community Development in Batanggas. University of The Philippines : A Community Development Research Counsil Publication.

[7] Sikhondze, Wilson B. 1999. "The Role of Extension in Farmer Education and Information Dissemination in Swaziland". Journal: Adult Education and Development No. 53/1999, Institute for International Cooperation of The German Adult Education Association, Bonn : 112/DW

[8] Korten, David C., 1980, Community Organization and Rural Development: A Learning Process Approach, Public Administration Review, September/October 1980 p.480509.

[9] Sumodiningrat, Gunawan. (1999) Community Empowerment and Social Security Nets. Yogyakarta, Ghalia Indonesia.

[10] Marmujiono, Slamet Priyo, (2014), Analysis of Factors Affecting Poverty Levels in Poverty Alleviation Strategies in Kab. Brebes 2009-2011 ", Jurnal Analisis Ekonomi Pembangunan, Vol.3, No.1, page 160-172.

[11] Central Bureau of Statistics. 2015. Sustainable Development Indicators. Jakarta: Central Bureau of Statistics

[12] Rokhim, (2014), "The Role of Muhammadiyah's Organization in Education in the District of Sukorejo, Kendal District", Jurnal Ilmu Pendidikan Sejarah, Vol.02, No.1, page 23-31.

[13] Muhammadiyah Central Executive Board. 2018. Characteristics of the Struggle of Muhammadiyah. http://www.muhammadiyah.or.id

[14] Muhammadiyah Central Management Community Empowerment Assembly. 2018. Scope and Role of the Community Empowerment Assembly. http://mpm.muhammadiyah.or.id

[15] Creswell, John W. 2009. Research Design ualitative, Quantitative, and Mixed Research Approaches. Yogyakarta: Pustaka Pelajar. Translator Achmad Fawaid.

An interval energy-water nexus planning model for electric power planning

D Zhao[1,2,3], Y P Li[12,3], J Liu1 and H Y Fu[1]

1 College of Environmental Science and Engineering, Xiamen University of Technology, Xiamen 361024, China
2 Key Laboratory of Environmental Biotechnology (XMUT), Fujian Province University
3 Fujian Engineering Research Center of Rural Sewage Treatment and Water Safety
E-mail: 414820675@qq.com, zyljing@126.com

Abstract. Reasonable energy planning plays a vital role in urban development and even the development of the whole national economy. Energy planning is an important way to solve the problems of resources and environment, promote energy conservation and emission reduction, and ensure the sustainable development of cities. Taking Xiamen City as an example, this paper uses the method of interval linear programming, aiming at minimizing the total cost of the system, and taking energy availability, water resources availability, energy demand and pollutant discharge as constraints, establishes an optimization scheme that conforms to the characteristics of Xiamen's energy system. The research results show that the clean energy utilization technology should be improved to further increase the development and utilization of local renewable energy, and diversify the power supply methods to achieve safe, economic, clean and good urban development.

1 Introduction

At present, Xiamen is in the stage of rapid economic development, and its development has a huge demand for energy. Rapid economic growth will inevitably drive the rapid growth of energy consumption. Effective planning and management of energy systems is particularly important. Compared with foreign countries, domestic research on urban energy system planning started relatively late, mainly focusing on the application of foreign energy models. For example, Liang used linear programming methods such as interval programming, fuzzy programming and stochastic programming to solve the uncertain power supply optimization model in Beijing, obtained the energy demand for power generation and the installed capacity expansion of generating units [1]. Taking the power grid of Jiangsu Province as an example, Du et al. established a low-carbon grid planning model adapted to low-carbon power generation based on multi-scenario modeling technology considering various low-carbon elements of the grid link. [2]. Zhou et al. constructed Shenzhen sustainable power planning model based on robust possibility planning method, obtained energy supply, power supply structure, outsourced power consumption scheme under the condition of minimizing power supply cost [3]. Yu et al. used MARKAL model to conduct scenario analysis of Shanghai energy system [4]. Jia et al used LEAP model to study the energy demand and carbon dioxide emissions of Jingdezhen, Jiangxi [5]. Zeng et al used linear programming to model and simulate the integrated power resource planning considering carbon emission constraints [6]. Feng et al used LEAP model to study the scenarios of energy conservation and carbon emission reduction policies in Beijing [7]. However, most of the domestic studies are limited to the application of foreign models, and lack of

Content from this work may be used under the terms of the Creative Commons Attribution 3.0 licence. Any further distribution of this work must maintain attribution to the author(s) and the title of the work, journal citation and DOI.
Published under licence by IOP Publishing Ltd

comprehensive consideration of the characteristics of China's energy system. The purpose of this study is to target the energy situation in Xiamen for a certain period of time, with the goal of minimizing the cost of energy systems, taking the urban power generation heating system as the core, considering energy consumption, energy supply and demand, energy conversion, pollutant discharge and other factors. It can provide some new ideas for decision makers on the planning of urban energy systems.

2 Study system description

2.1 Study area

Xiamen is located on the southeastern coast of China, in the southern part of Fujian Province. It is inland bordering Zhangzhou and Quanzhou. In the east, it is separated from Jinmen Island and Longhai City in size. It is located at 24°23'~24°54' north latitude and 117°53'~118°26' east longitude. The total area of Xiamen is 1699.39 km^2, which is composed of six municipal districts as shown in Figure 1.

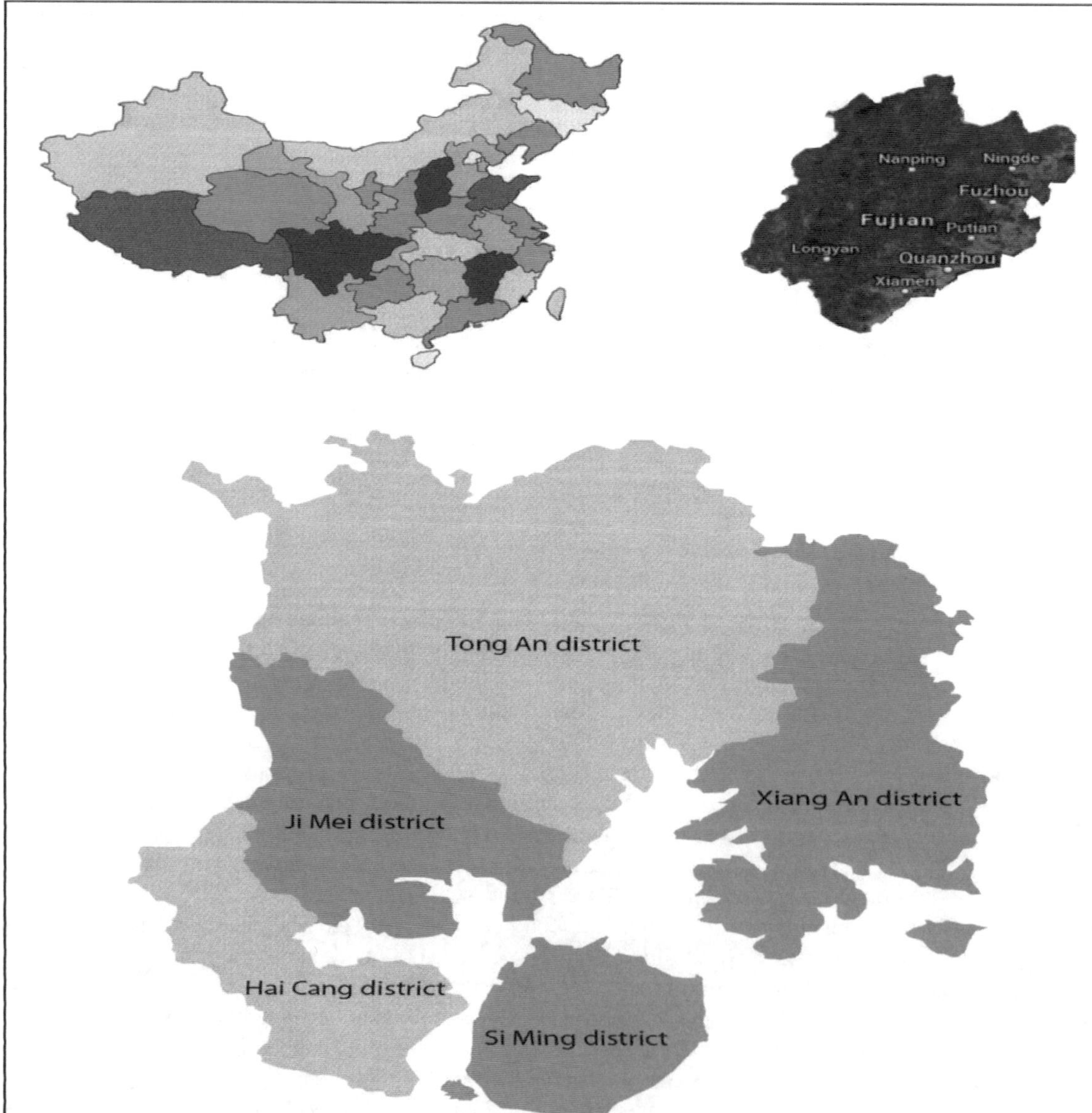

Figure 1. Geographical position of Xiamen.

2.2 Energy flow

Xiamen is a typical energy-consuming city with 99% of its energy supply depending on imports. According to statistics, in the energy purchases of Industrial Enterprises above the scale in 2017, raw coal purchased 3.32 million tons, natural gas purchased 345.75 million m^3, and power purchased from abroad reached 8633.01 million kWh. The purchased power is directly imported into the power grid and transmitted to various user terminals. The primary energy purchased is transformed by the processing of power plants and thermal power plants, and the converted power and heat are input into the urban power grid and heat network system, and eventually supplied to the end-users. During the process of energy conversion, SO_2, NO_X and PM10 discharged from power plants and thermal power plants will pollute the urban environment, which needs to be controlled and treated.

2.3 Problem description

The energy consumption structure of Xiamen City is not healthy, mainly based on non-renewable energy consumption such as coal and natural gas. In 2017, coal consumption accounted for more than 50% of the regular energy consumption. In order to better protect the city's ecological environment, it is necessary to change its energy consumption structure, and more clean and renewable energy generation technologies such as wind and solar energy should be developed and used. Considering the huge demand for energy in cities, in the transitional stage of energy structure transformation, people should improve energy processing and conversion technology to achieve the purpose of improving energy conversion efficiency. The planning of urban energy system needs to take into account such factors as energy price, energy supply and demand, energy processing and conversion, energy structure and pollutant emission control. There are also a lot of uncertainties in energy system planning, the most important one is the uncertainty of data parameters. These complexities and uncertainties affect the availability of models in the process of energy system planning. In order to solve the above-mentioned problems of complexity and uncertainty, this study adopts the method of interval linear programming to deal with the uncertainty of data parameters in the way of interval expression of upper and lower bounds.

3. Modelling development

3.1Model Construction

Interval linear programming [8] is a typical optimization analytical method for planning under uncertainty. It provides satisfactory solutions by allowing uncertainty to be effectively reflected in coefficients and objective functions without distribution functions. Interval linear programming proposed by Huang et al is widely used in dealing with uncertain information of energy system planning.

In this study, the planning period of the model is 15 years and five years is a planning period divided into three periods.

Objective function:

$$Minf = (a) + (b) + (c) + (d) + (e) + (g) + (h) \tag{3.1}$$

In the formula, f is the total cost of the system, a is the purchase cost of energy, b is the cost of power generation, c is the cost of heating, d is the cost of expansion, e is the cost of power transmission, g is the cost of pollutant control, and h is the cost of water resources utilization.

(1) Energy purchase cost:

$$\sum_{i=1}^{3}\sum_{k=1}^{3} ZN_{i,k}^{\pm} \times LN_{i,k}^{\pm} \tag{3.2}$$

In the formula, i represents the type of energy purchases, i = 1, 2, 3 represents raw coal, natural gas and electricity, respectively. K is the planning period, k = 1, 2, 3. $LN_{i,k}^{\pm}$represents the purchase of energy type i in the k period. $ZN_{i,k}^{\pm}$represents the purchase cost of energy type i in k period.

(2) Generation costs:

$$\sum_{n=1}^{7}\sum_{k=1}^{3}\left(PEG_{n,k}^{\pm}\times PDR_{n,k}^{\pm}\right)+\sum_{n=8}^{9}\sum_{k=1}^{3}\left(PRR_{n,k}^{\pm}\times PRC_{n,k}^{\pm}\right) \tag{3.3}$$

In the formula, n stands for the type of energy conversion technology, n = 1, 2, 3, 4, 5, 6, 7, 8, 9 for coal-fired power generation, gas-fired power generation, biomass power generation, pumped storage power generation, hydroelectric power generation, photovoltaic power generation, wind power generation, coal-fired cogeneration and gas-fired cogeneration. $PEG_{n,k}^{\pm}$ represents the unit generation cost of power conversion technology type n of power plant in k period. $PDR_{n,k}^{\pm}$ represents the power generation of power plant conversion technology type n in k period. $PRR_{n,k}^{\pm}$ represents the power generation of the conversion technology type n of thermal power plant in k period. $PRC_{n,k}^{\pm}$ represents the unit cost of power generation for conversion technology type n of thermal power plant in k period.

(3) Heating cost:

$$\sum_{n=8}^{9}\sum_{k=1}^{3}\left(HRR_{n,k}^{\pm}\times HRC_{n,k}^{\pm}\right) \tag{3.4}$$

In the formula, $HRR_{n,k}^{\pm}$ represents the heat transfer technology type n of the thermal power plant in the k period, and $HRC_{n,k}^{\pm}$ represents the heat transfer operation cost of the technology type n of the thermal power plant in the k period.

(4) Expansion cost:

$$\sum_{n=1}^{7}\sum_{k=1}^{3}\left(PDA_{n,k}^{\pm}\times PDI_{n,k}^{\pm}\right)+\sum_{n=8}^{9}\sum_{k=1}^{3}\left(PRA_{n,k}^{\pm}\times PRI_{n,k}^{\pm}\right) \tag{3.5}$$

In the formula, $PDA_{n,k}^{\pm}$ represents the capacity expansion of conversion technology n in different expansion plans of power plants in k period. $PDI_{n,k}^{\pm}$ represents the investment cost of the conversion technology type n of power plant in k period. $PRA_{n,k}^{\pm}$ represents the capacity expansion of down-conversion technology n in different expansion plans of thermal power plants in k period. $PRI_{n,k}^{\pm}$ represents the investment cost of conversion technology type n of thermal power plant in k period.

(5) Power transmission cost:

$$\sum_{k=1}^{3}\left(\sum_{n=1}^{7}PDR_{n,k}^{\pm}+\sum_{n=8}^{9}PRR_{n,k}^{\pm}\right)\times Cu_{k}^{\pm}$$
$$+\sum_{k=1}^{3}\left(\sum_{n=1}^{7}PDA+\sum_{n=8}^{9}PRA\times\left(1-\beta_{n,k}^{\pm}\right)\times h_{n,k}^{\pm}\right)\times Cu_{k}^{\pm} \tag{3.6}$$

In the formula, $Cu_{n,k}^{\pm}$ represents the hourly transportation cost of electricity in k period. $h_{n,k}^{\pm}$ stands for hours of power generation in k period. $\beta_{n,k}^{\pm}$ represents the ratio of heat and power production capacity of thermal power plant in k period.

(6) Pollutant control costs:

$$\sum_{n=1}^{7}\sum_{k=1}^{3}\sum_{e=1}^{3}\left(PDR_{n,k}^{\pm}+PDA_{n,k}^{\pm}\times h_{n,k}^{\pm}\right)\times 3.6\times CE_{e,k}^{\pm}$$
$$+\sum_{n=8}^{9}\sum_{k=1}^{3}\sum_{e=1}^{3}\left[PRR_{n,k}^{\pm}+\left(HRR_{n,k}^{\pm}+PRA_{n,k}^{\pm}\times h_{n,k}^{\pm}\right)\times 3.6\right]\times CE_{e,k}^{\pm} \tag{3.7}$$

In the formula, e represents the type of energy pollution, e = 1, 2, 3 represents SO_2, NOx, PM10, respectively. $CE_{n,k}^{\pm}$ represents the emission cost of pollution type e in k period.

(7) Cost of water resources utilization:

$$\sum_{r=1}^{3}\sum_{k=1}^{3}CW_{k}^{\pm}\times\left(VWE_{r,k}^{\pm}+VWH_{r,k}^{\pm}\right) \tag{3.8}$$

In the formula, r represents the type of water resources, and r = 1, 2 and 3 respectively represents surface water, groundwater and reused water. $CW^{\pm}_{n,k}$ represents the cost of r-type water resources supplied by power plants during the k-period. $VWE^{\pm}_{n,k}$ represents the amount of R-type water resources supplied by power plants in k period, and $VWH^{\pm}_{n,k}$ represents the amount of r-type water resources supplied by thermal power plants in k period.

The constraints are as follows:

a. Energy availability:

$$LN^{\pm}_{i,k} \le ZO^{\pm}_{i,k}, \forall i,k \tag{3.9}$$

In the formula, $ZO^{\pm}_{i,k}$ is the maximum available amount the maximum available amount of energy type i in k period

b. Availability of water resources:

$$\begin{aligned} &VWE^{\pm}_{r,k} \le AWE^{\pm}_{r,k}, \forall r,k \\ &VWH^{\pm}_{r,k} \le AWH^{\pm}_{r,k}, \forall r,k \end{aligned} \tag{3.10}$$

In the formula, $AWE^{\pm}_{r,k}$ is the maximum available amount of water resource type r used by power plants in k period. $AWH^{\pm}_{r,k}$ is the maximum available amount of water resource type r used by thermal power plants in k period.

c. Energy and material balance constraints:

$$\begin{aligned} &\sum_{n=1}^{7}\left(PDR^{\pm}_{n,k} + PDA^{\pm}_{n,k} \times h^{\pm}_{n,k}\right) \times FE^{\pm}_{n,k} \\ &+\sum_{n=8}^{9}\left(PRR^{\pm}_{n,k} + PRA^{\pm}_{n,k} \times \left(1-\beta^{\pm}_{n,k}\right) \times h^{\pm}_{n,k}\right) \times FE^{\pm}_{n,k} \\ &+\sum_{n=8}^{9}\left(HRR^{\pm}_{n,k} + PRA^{\pm}_{n,k} \times \beta^{\pm}_{n,k} \times h^{\pm}_{n,k}\right) \times HE^{\pm}_{n,k} \le \sum_{i=1}^{3} XN^{\pm}_{i,k}, \forall k \end{aligned} \tag{3.11}$$

In the formula, $FE^{\pm}_{n,k}$ represents the energy consumption per unit of power generation in k period. $HE^{\pm}_{n,k}$ r is the energy consumption per unit of n energy type in k period.

d. Energy demand constraints:

$$\begin{aligned} &\sum_{n=8}^{9}\left(HRR^{\pm}_{n,k} + PRA^{\pm}_{n,k} \times \beta^{\pm}_{n,k} \times h^{\pm}_{n,k}\right) \ge DH^{\pm}_{k}, \forall k \\ &\sum_{n=1}^{7}\left(PDR^{\pm}_{n,k} + PDA^{\pm}_{n,k} \times h^{\pm}_{n,k}\right) \\ &+\sum_{n=1}^{7}\left(PRR^{\pm}_{n,k} + PRA^{\pm}_{n,k} \times \left(1-\beta^{\pm}_{n,k}\right) \times h^{\pm}_{n,k}\right) \ge DE^{\pm}_{k}, \forall k \end{aligned} \tag{3.12}$$

In the formula, $DH^{\pm}_{k}$ r is the minimum local heat demand in k period. $DE^{\pm}_{k}$ is the local minimum electricity demand in k period.

e. Energy-water resource linkage constraints:

$$\sum_{n=8}^{9}\left[HRR^{\pm}_{n,k} + PRA^{\pm}_{n,k} \times \beta^{\pm}_{n,k} \times h^{\pm}_{n,k}\right] \times UWH^{\pm}_{n,k} \ge DHW^{\pm}_{k}, \forall k \tag{3.13}$$

$$\begin{aligned} &\sum_{n=1}^{7}\left(PDR^{\pm}_{n,k} + PDA^{\pm}_{n,k} \times h^{\pm}_{n,k}\right) \times UWE^{\pm}_{n,k} \\ &+\sum_{n=1}^{7}\left(PRR^{\pm}_{n,k} + PRA^{\pm}_{n,k} \times \left(1-\beta^{\pm}_{n,k}\right) \times h^{\pm}_{n,k}\right) \times UWE^{\pm}_{n,k} \ge DEW^{\pm}_{k}, \forall k \end{aligned} \tag{3.14}$$

$$\sum_{n=8}^{9}\left(HRR^{\pm}_{n,k} + PRA^{\pm}_{n,k} \times \beta^{\pm}_{n,k} \times h^{\pm}_{n,k}\right) \times UWH^{\pm}_{n,k} \le \sum_{r=1}^{3} VWH^{\pm}_{r,k}, \forall k \tag{3.15}$$

$$\left(\sum_{n=1}^{7} PDR_{n,k}^{\pm} + \sum_{n=1}^{7} PDA_{n,k}^{\pm} \times h_{n,k}^{\pm}\right) \times UWE_{n,k}^{\pm}$$

$$+\sum_{n=8}^{9}\left(PRR_{n,k}^{\pm} + PRA_{n,k}^{\pm} \times \left(1-\beta_{n,k}^{\pm}\right) \times h_{n,k}^{\pm}\right) \times UWE_{n,k}^{\pm} \le \sum_{r=1}^{3} VWE_{r,k}^{\pm}, \forall k \quad (3.16)$$

$$\sum_{r=1}^{3} VWH_{r,k}^{\pm} \times UHW_{k}^{\pm} \le \sum_{n=8}^{9}\left(HRR_{n,k}^{\pm} + PRA_{n,k}^{\pm} \times \beta_{n,k}^{\pm} \times h_{n,k}^{\pm}\right) \times \delta_{n,k}^{\pm}, \forall k \quad (3.17)$$

$$\sum_{r=1}^{3} VWE_{r,k}^{\pm} \times UEW_{k}^{\pm} \le \left(\sum_{n=1}^{7} PDR_{n,k}^{\pm} + \sum_{n=1}^{7} PDA_{n,k}^{\pm} \times h_{n,k}^{\pm}\right) \times \theta_{n,k}^{\pm}$$

$$+\sum_{n=8}^{9}\left(PRR_{n,k}^{\pm} + PRA_{n,k}^{\pm} \times \left(1-\beta_{n,k}^{\pm}\right) \times h_{n,k}^{\pm}\right) \times \theta_{n,k}^{\pm}, \forall k \quad (3.18)$$

f. Pollutant emission control constraints:

$$\sum_{i=1}^{3} LN_{i,k}^{\pm} \times EP_{i,k}^{\pm} \times \left(1-SF_{e,k}^{\pm}\right) \le ES_{e,k}^{\pm}, \forall e,k \quad (3.19)$$

g. Non-negative constraints:

$$LN_{i,k}^{\pm}, PDR_{n,k}^{\pm}, PRR_{n,k}^{\pm}, HRR_{n,k}^{\pm}, PDA_{n,k}^{\pm}, PRA_{n,k}^{\pm}, VWE_{r,k}^{\pm}, VWH_{r,k}^{\pm} \ge 0 \quad (3.20)$$

3.2Data collection

The data were collected from Xiamen Statistical Yearbook, Government Departments Development Planning, Relevant Literature and Case Studies, etc. Table 1 shows more details of the energy consumption of Xiamen City from 2011 to 2015.

Table 1. Energy consumption of Xiamen.

Year	electricity consumption/ 10^8 kWh	Natural-gas consumption 10^4 m^3	Coal consumption 10^4 ton	Water consumption $10^4 m^3$
2011	163.2	56466	513.4	43566
2012	174.8	52240	404.7	45244
2013	195.8	54246	485.6	48632
2014	209.3	49170	499.2	51781
2015	211.4	33200	435.5	52255

4. Results analysis and discussion

4.1 Electric power generation

The model is solved by the interval linear programming method. The operation yields the result shown in Figure 2. From the planning period k=1 to k=3, the coal-fired power generation is [42.33, 44.68]× 10^6 MW, [41.61, 43.20] × 10^6MW and [40.44, 42.90] × 10^6 MW. The power generation is declining, but overall, coal-fired power generation still provides the vast majority of electricity and occupies the dominant position. Compared with other power generation technologies, the power generation capacity of biomass power generation, wind power generation, photovoltaic power generation and coal-fired gas-fired cogeneration technology have been improved rapidly, indicating that they have good development potential. The future energy development direction of Xiamen should vigorously promote the application of biomass, wind, photovoltaic and coal-fired gas-fired cogeneration technology, which is conducive to the sustainable development of urban energy.

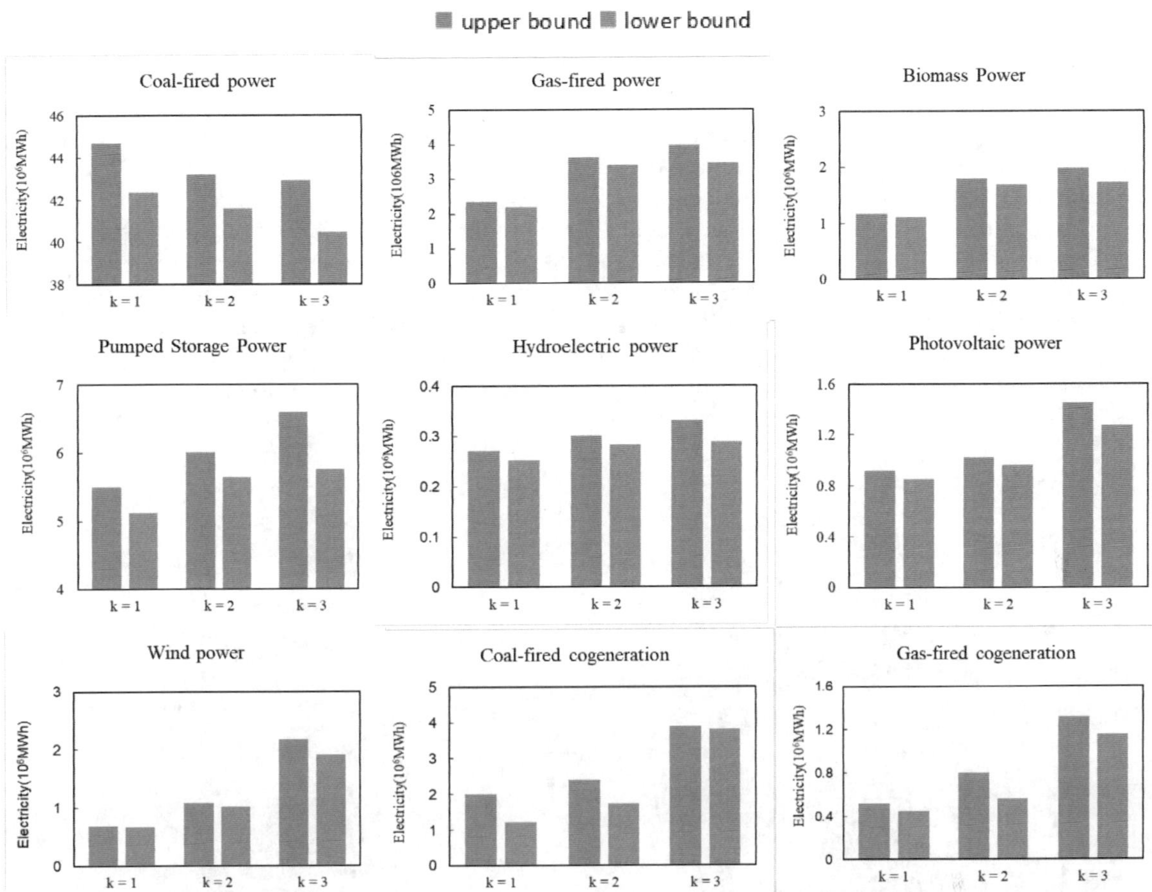

Figure 2. Generation capacity of different generation technologies in each planning period.

4.2 Analysis of heat supply

Coal-fired cogeneration and gas-fired cogeneration are the main heating modes in Xiamen, as shown in Figure 3. Coal-fired cogeneration is [37.4,39.4]×10^6 GJ at planning period k=1, much larger than gas-fired cogeneration [6.1,6.2]×10^6 GJ; coal-fired cogeneration is [35.9,37.8]×10^6 GJ at planning period k=2, gas-fired cogeneration is [6.4,6.7]×10^6 GJ; gas-fired cogeneration is [6.6,6.9]×10^6 GJ at planning period k=3, and coal-fired cogeneration is [33.0,36.8]×10^6 GJ.

The overall development trend is to reduce coal-fired heating and increase the heat supply of gas. For Xiamen, vigorously developing gas and reducing coal-fired heating are of great significance to the protection of urban environment.

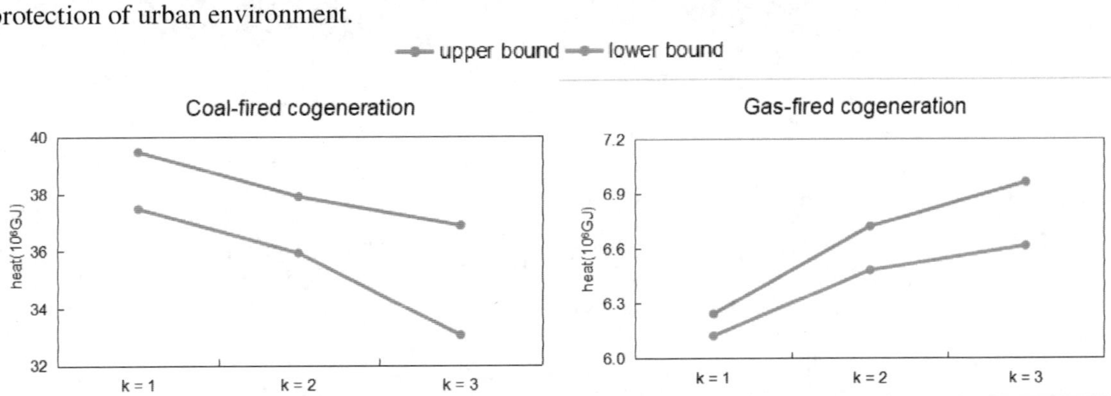

Figure 3. Heat production of two heating modes in different periods.

IOP Conf. Series: Earth and Environmental Science **356** (2019) 012003 doi:10.1088/1755-1315/356/1/012003

4.3 Expansion analysis

The expansion of Xiamen's electricity capacity is shown in Figure 4. According to the urban development trend of Xiamen, considering the expansion of coal-fired power generation to zero, it has certain practical significance for reducing coal-fired pollution emissions. Considering the shortage of hydropower resources in Xiamen and the saturation of its installed capacity, the expansion capacity of Xiamen is set to zero. Gas-fired cogeneration and wind power expansion account for a larger proportion in the three planning periods, especially the proportion of wind power continues to increase. The diagram shows that the energy structure of Xiamen has been improved, photovoltaic power generation and wind power generation have been developed efficiently, and cogeneration projects have been greatly improved. In the future, Xiamen's energy development strategy will still be to improve the utilization of clean and renewable energy, reduce coal-fired power generation or transform into cogeneration projects, which will contribute to the city's environmental protection.

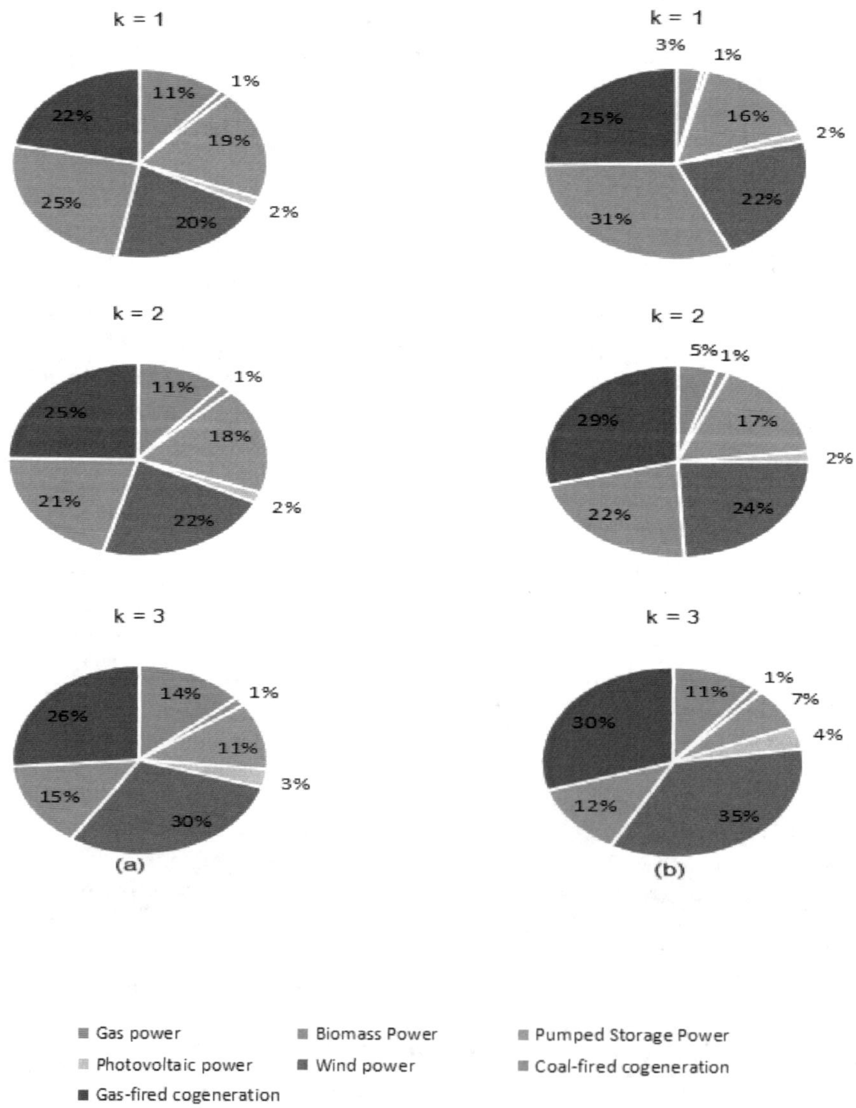

Figure 4. (a) The upper limit of each generation technology expansion in three periods and (b) the proportion of the lower limit of each generation technology expansion in three periods.

4.4 Water consumption analysis

As shown in Figure 5, the water consumption of Xiamen's thermal and power industry has remained basically unchanged from k=1 to k=3 planning period; the surface water consumption of power plants has decreased from [1439,1998]×10^6m^3, [1200,1607]×10^6m^3 to [1127,1477]×10^6m^3, maintaining an average decline, the proportion of repetitive water consumption has maintained a certain level, and the reuse rate of groundwater resources has maintained a relatively high level, while the utilization rate of groundwater resources has been maintained. The decline trend of water quantity is obvious, which may be related to Xiamen's restriction policy on groundwater exploitation. In the future, Xiamen's water resources management should pay more attention to the reuse of reused water and the sustainable development benefits of water resources.

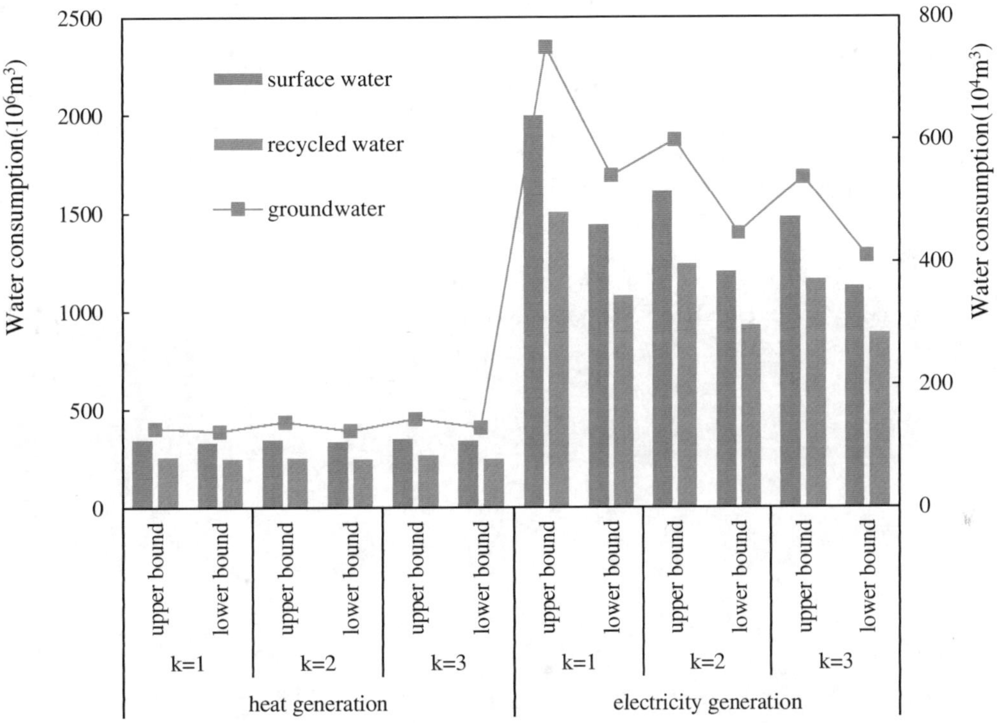

Figure 5. Water consumption of heat generation and electricity generation.

5. Discussion

At present, Xiamen coal-fired power generation still occupies a dominant position. Gas-fired power generation, biomass power generation, pumped storage power generation, hydroelectric power generation, photovoltaic power generation, wind power generation, coal-fired cogeneration and gas-fired cogeneration are supplementary. With the increasing proportion of renewable energy generation, the traditional fossil energy generation will gradually decline, and the renewable energy generation will generally show an upward trend.

Xiamen's coal-fired heat supply is gradually decreasing, while the gas-fired heat supply is gradually increasing, which is related to Xiamen's efforts to develop gas-fired cogeneration projects and gradually reduce the use of coal. In terms of water use, the water consumption of thermal power plants is basically unchanged, and the water consumption of power plants gradually decreases, which is also related to the gradual reduction of traditional fossil energy consumption.

IOP Conf. Series: Earth and Environmental Science **356** (2019) 012003 doi:10.1088/1755-1315/356/1/012003

6. Reference

[1] Liang Y X 2009 *Urban energy optimization model under uncertain conditions* (Bei Jing: North China Electric Power University) p 1-72
[2] Du E S, Sun Y L and Zhang N 2015 Low-carbon power grid planning model for low-carbon power supply development *Power grid technology.* **39** 2725–30
[3] Zhou Y, Li Y P and Huang G H 2014 Shenzhen sustainable electricity planning Model and energy conservation and emission reduction analysis *Electricity Construction.* **35** 75-80
[4] Yu Y F, Hu J Y and Zhang S R 2008 MARKAL model and scenario analysis of Shanghai energy system *Journal of Shanghai Jiao Tong University.* **42** 360–69
[5] Jia Y P and Liu R Z 2010 Urban energy planning and CO_2 emission reduction based on LEAP model: A case study of Jingdezhen *Journal of Applied Basic and Engineering Sciences* **18** 75-83.
[6] Zeng M[, Qin X L and Ma X C 2010 *East China Electric Power.* **38** 1103–7
[7] Feng Y Y, less tension.2012 Scenario analysis of urban energy conservation and carbon emission reduction policies: Taking Beijing as an example *Resource Science.* **34** 541–50.
[8] Huang G H and Moore R D.1993 Grey linear programming, its solving approach, and its application. *Int J Syst Sci.* **24** 159–72

Acknowledgements

This research was supported by Fujian Engineering Research Center of Rural Sewage Treatment and Water Safety and High-level Talent Program of Xiamen University of Technology (YKJ17018R).

Effect of heat treatment on hydrogen production by combined fermentation of wild carp and excess sludge

H D Zhu[1], H Y Fu[1*], M Xu[2], G X Su[3], Y C Wu[1], P F Gao[1], L Jin[1], S Lin[1]

1 Key Laboratory of Environmental Biotechnology(XMUT), Fujian Province University, Xiamen University of Technology, Xiamen 361024, China
2 Chinese Academy for Environmental Planning, Beijing 100012, China
3 Collaborative Innovation Center for Marine Biotechnology, Higher Vocational College, Fujian Province, Xiamen Ocean Vocational College, Xiamen 361100, China
E-mail: fuhy@xmut.edu.cn

Abstract. a lignocellulose-rich biomass, the wild carp can be efficientlyhydrolyzed into reducing sugar after a certain degree of pretreatment, and it is used as a high-quality raw material for the development and utilization of biomass energy such as fuel ethanol and hydrogen. In this study, the wild carp was pretreated at 50 °C, 60 °C, 70 °C, 80 °C, and 90 °C to investigate the effects of different heat treatment temperatures on the hydrogen production of wild mash and excess sludge mixed fermentation, and analyzed by gas chromatography. The experimental results show that when the pretreatment temperature is 70 °C, the gas production is the highest, 216mL, the highest hydrogen production is 43.01mL, and the hydrogen concentration is up to 20.73%.

1. Introduction

With the intensification of the energy crisis, countries around the world have accelerated the development of biomass energy [1-2], including bio-hydrogen, biogas, fuel ethanol and so on. As a renewable energy source, lignocellulose is hydrolyzed to produce reducing sugars such as glucose and xylose, and anaerobic fermentation of reducing sugars can produce biohydrogen. However, the binding of lignin components to cellulose causes a decrease in reducing sugar production, which affects fermentative hydrogen production. Therefore, the substrate needs to be pretreated before fermentation to remove the lignin from the surface of cellulose and hemicellulose to achieve the effect of removing lignin [3-8]. The wild carp is rich in sugar and fiber materials, which can be used for fermentative hydrogen production. The anaerobic fermentation gas production technology can realize the recycling of resources and energy in the wild carp. The excess sludge contains a large amount of microorganisms and is rich in organic matter such as protein and polysaccharide[9]. Therefore, in this experiment, the excess sludge and wild carp are used as substrates to study the hydrogen production test after pretreatment of wild carp at different temperatures, in order to obtain wild carp. The best heat treatment method provides certain technical support for the utilization of

Content from this work may be used under the terms of the Creative Commons Attribution 3.0 licence. Any further distribution of this work must maintain attribution to the author(s) and the title of the work, journal citation and DOI.
Published under licence by IOP Publishing Ltd

biomass energy in the wild donkey.

2. Experimental materials and methods

2.1 Experimental materials

The experimental wild donkey was collected from a village in Xiamen. The remaining sludge was taken from a sewage treatment plant in Xiamen. The sludge was stored at 4 °C, and the residual sludge solids ratio (TS) was about 3%.

2.2 Wild carp pretreatment method

The dried wild cockroaches were dissolved in 30 mL of distilled water and placed in a constant temperature water bath at 50 ° C, 60 ° C, 70 ° C, 80 ° C, and 90 ° C, water bath for 30 min, and then cooled to room temperature under natural conditions.

2.3 Hydrogen production test of combined sludge and wild carp

The pre-treated residual sludge and wild carp fermentation hydrogen production test was carried out in a 250 mL Erlenmeyer flask, and the pH of the substrate was adjusted to neutral before the sludge fermentation. Take 250 mL of excess sludge and an appropriate amount of wild scorpion in an Erlenmeyer flask, and purge the Erlenmeyer flask with nitrogen for 5 min to remove excess oxygen and quickly seal the device. The sealed conical flask was placed in a constant temperature water bath and anaerobic fermentation was carried out at 37 °C. The fermentation gas was collected and analyzed by the drainage method. The experimental setup is shown in Figure 1.

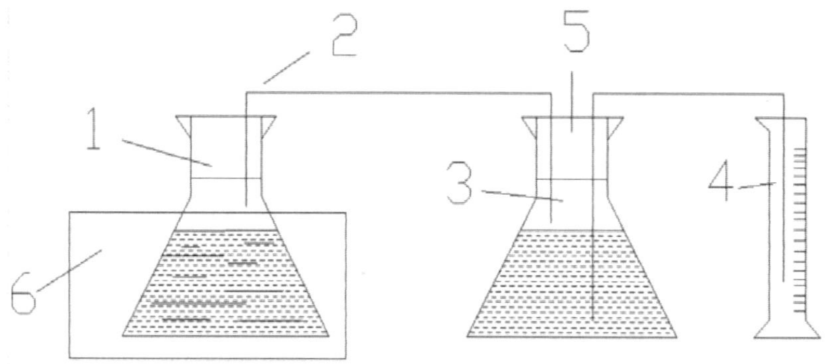

Figure 1. Experimental device schematic

1-fermented bottle; 2-linked glass elbow; 3-drainage bottle and gas collection device; 4-measuring cylinder; 5-sampling port; 6-constant water bath

2.4 Gas Chromatograph Composition Analysis of Fermentation Gas

The composition of the fermentation gas was analyzed by Agilent Technologies 7890B GC System gas chromatograph. The column type was 5A and the diameter was 3mm × height 2mm. The parameter setting was: column temperature 50 °C, injector temperature 50 °C, thermal conductivity detector The temperature was 120 ° C, the carrier gas was helium, and the flow rate was 30 mL/min. The concentration of hydrogen gas was measured.

3. Experimental results and analysis

3.1 Effect of heat treatment method on hydrogen production from wild carp fermentation

Pretreatment of wild carp was carried out at different temperatures of 50 ° C, 60 ° C, 70 ° C, 80 ° C, and 90 ° C to investigate the effect of temperature on gas production. The experimental results are shown in Table 1.

Table 1. Effect of different temperature on hydrogen production by combined fermentation of wild carp and excess sludge [not a three-line table]

temperature(°C)	gas production（mL）
50	156.5
60	175.5
70	216
80	175.5
90	171

It can be seen from Table 1 that when the temperature is 50 ° C, the gas production is 156.5 mL, the gas production is 175.5 mL when the temperature is 60 ° C, the gas production is 216 mL when the temperature is 70 ° C, and the gas production is 80 ° C. 175.5 mL; when the temperature is 90 ° C, the gas production is 171 mL. The gas production increases with the increase of temperature. When the temperature is 70 °C, the gas production reaches the highest. As the temperature continues to rise, the gas production begins to decrease.

3.2 Gas chromatograph analysis of the composition of the fermentation gas

The composition of the fermentation gas was analyzed by gas chromatography to study the effect of different temperatures on the hydrogen production after pretreatment. The experimental results are shown in Figures 2, 3, 4, 5 and 6.

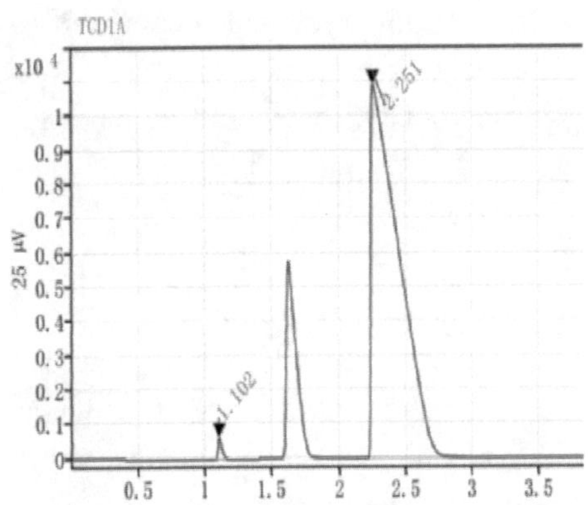

Figure 2. Gas composition analysis at 50 °C

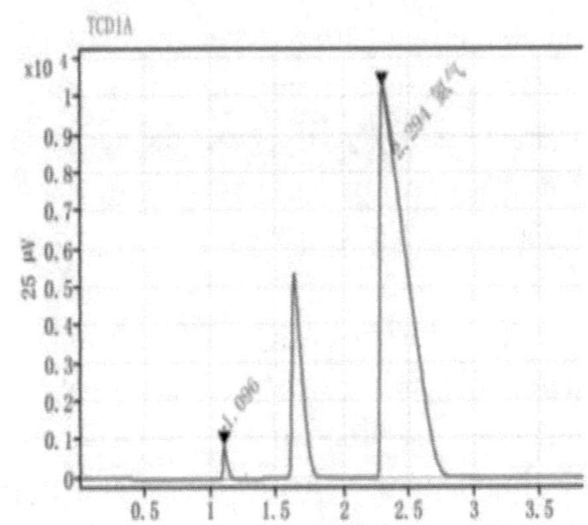

Figure 3. Gas composition analysis at 60 °C

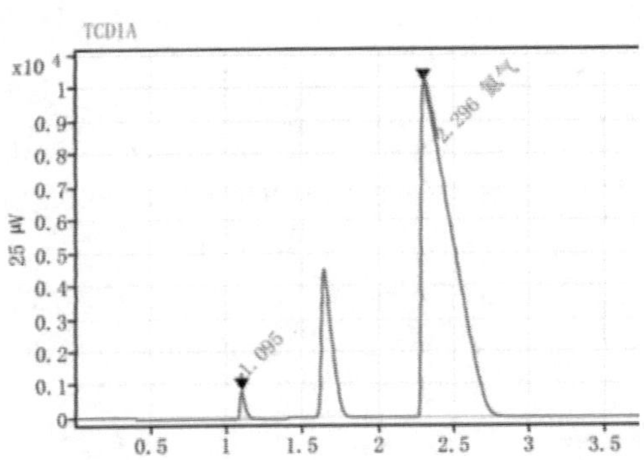

Figure 4. Gas composition analysis at 70 °C

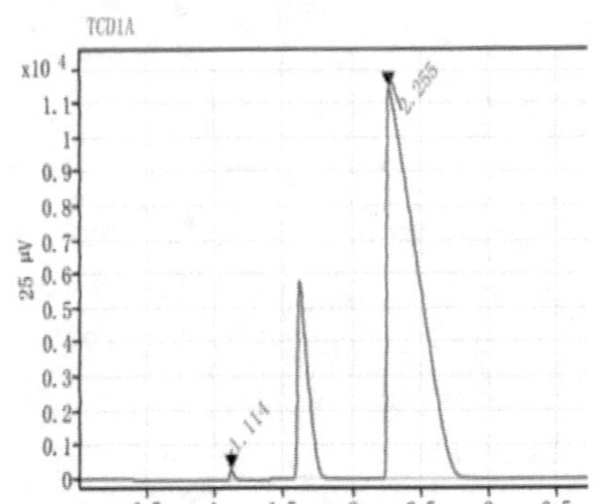

Figure 5. Gas composition analysis at 80 °C

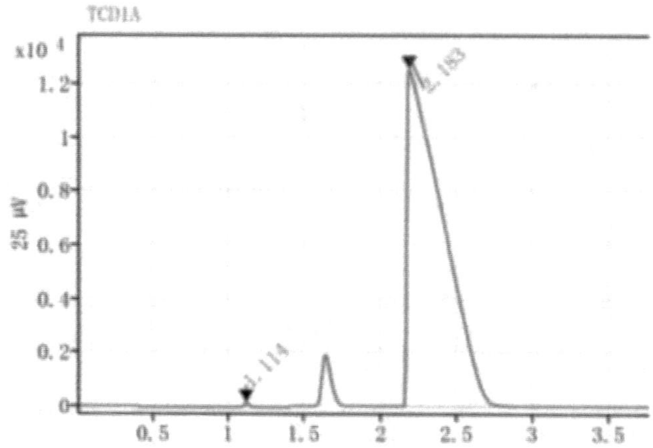

Figure 6. Gas composition analysis at 90 ° C

The experimental results show that when the temperature is 50 °C, the hydrogen concentration is 13.867%, the hydrogen production is 21.7mL; when the temperature is 60 °C, the hydrogen concentration is 16.857%, the hydrogen production is 29.58mL; the temperature is 70 °C, the hydrogen The concentration was 20.73%, the hydrogen production was 43.01mL, the hydrogen concentration was 15.441%, the hydrogen production was 27.10mL, the hydrogen concentration was 13.992%, and the hydrogen production was 23.93mL. With the increase of temperature, the activity of hydrogen-producing bacteria gradually increases. When the temperature is 70 °C, the activity of hydrogen-producing bacteria is the best. As the temperature continues to rise, the high temperature destroys the structure of hydrogen-producing bacteria and denatures the hydrogen-producing bacteria. Thereby affecting the amount of hydrogen produced.

4. Conclusion [The conclusion is best to have two points, write an introduction in front

(1) The experimental results show that different temperatures have a certain effect on the hydrogen production by co-pretreatment with residual sludge fermentation. With the increase of temperature, the gas production gradually increases. When the temperature reaches 70 °C, the gas production reaches the highest. At 216 mL, as the temperature continues to rise, the gas production begins to decrease.

(2) Analysis of the fermentation gas by gas chromatograph shows that the hydrogen concentration increases with the increase of temperature. When the temperature reaches 70 °C, the hydrogen concentration reaches 21.73%.

5. References

[1] Mehdi Mehrpooya,Maryam Khalili,Mohammad Mehdi Moftakhari Sharifzadeh. Model development and energy and exergy analysis of the biomass gasification process (Based on the various biomass sources)[J]. Renewable and Sustainable Energy Reviews,2018,91.

[2] Carlos Rodríguez-Monroy,Gloria Mármol-Acitores,Gabriel Nilsson-Cifuentes. Electricity generation in Chile using non-conventional renewable energy sources-A focus on biomass[J]. Renewable and Sustainable Energy Reviews,2018,81.

[3] Parameswaran B, Karri S, Raveendran S, et al. Short duration microwave assisted pretreatment enhances the enzymatic saccharification and fermentable sugar yield from sugarcane bagasse[J]. Renewable Energy, 2012,37:109-116.

[4] Raveendran S, Mathiyazhakan K, Parameswaran B, et al. Dilute acid pretreatment and enzymatic saccharification of sugarcane tops for bioethanol production[J].Bioresource Tec-
i. hnology, 2011, 102:10915-10921.

[5] Yadhu N G, Joelle D C, Florence H, et al. Comparison of some new pretreatment methods for second generation bioethanol production from wheat straw and water hyacinth [J]. Bioresource Technology, 2011, 102:4416-4424.

[6] Anita S, Narsi R B. Enzymatic hydrolysis optimization of microwave alkali pretreated wheat straw and ethanol production by yeast[J]. Bioresource Technology, 2012, 108:94-101.

[7] Xia Ao, Cheng Jun, Song Wenlu, et al. Enhancing enzymatic saccharification of water hyacinth through microwave heating with dilute acid pretreatment for biomass energy utilization [J]. Energy, 2013, 61:158-166.

[8] Paula A P, Albino A D, Irene F, et al. Influence of ligninolytic enzymes on straw saccharification during fungal pretreatment [J]. Bioresource Technology, 2012, 111:261-267.

[9] Guo L , Liu H , Li X , et al. [Bioproduction of volatile fatty acids from excess municipal sludge by multistage countercurrent fermentation].figJ]. Chinese Journal of Biotechnology, 2008, 24(7):1233-1239.

Acknowledgments

This work was financially supported by Fujian Science and Technology Guiding Project (2018Y0079), National Natural Science Foundation of China(51109181), Natural Science Foundation Project of Fujian Province(2018J01527).

IOP Conf. Series: Earth and Environmental Science **356** (2019) 012005 doi:10.1088/1755-1315/356/1/012005

Community Power Monitoring and Management System Based on LabVIEW

Zehong Huang[1],Guanlu Yang[2], Shuyang Wang[3],Xiaofeng Chen[4]

College of Information and Science Engineering,Huaqiao University, Xiamen , China
E-mail:535830236@qq.com

Abstract. With the wide application of power equipment and non-linear equipment in the community, the power quality of the power grid is further reduced. Therefore, power quality has become the research focus of the power industry today. This paper designs a community power monitoring and management system based on LabVIEW. The system consists of two parts: data acquisition circuit and software program. The specific functions include the communication configuration of LabVIEW and data acquisition card and the design of power quality monitoring. The feasibility of the power monitoring and management system.

1. Calculation and analysis of electrical energy indicators

1.1 Calculation and analysis of power supply voltage deviation

The supply voltage deviation refers to the difference between the actual measured voltage rms value and the system rated voltage value. The final voltage deviation value is expressed as a percentage.The calculation formula is:

$$\Delta U = \frac{U - U_N}{U_N} \times 100\% \qquad (1)$$

In the above formula, U is the measured voltage and U_N is the system nominal voltage. Voltage deviation is one of the most important indicators to pay attention to when studying power quality. In power systems, system reactive power is insufficient, power transmission distance is too long, system reactive power compensation is excessive, system load is too light or too heavy. Can cause deviation of the supply voltage.Among the above reasons, the lack of reactive power in the power grid is the most important cause of voltage deviation in the network[1].

1.2 Calculation and analysis of frequency deviation

The frequency deviation refers to the difference between the actually measured frequency and the rated frequency.Calculation formula:

$$\Delta f = f - f_0 \qquad (2)$$

In the formula, f is the measured frequency and f_0 is the rated frequency.

Frequency deviation is an important indicator of power quality. Whether the power system operates at high or low frequencies will have a great impact on power companies and power plants. Changes in frequency will cause unstable speeds in various motors, which will affect the quality of the products

Content from this work may be used under the terms of the Creative Commons Attribution 3.0 licence. Any further distribution of this work must maintain attribution to the author(s) and the title of the work, journal citation and DOI.
Published under licence by IOP Publishing Ltd

produced by the company. Due to the large number of impact loads and rectification equipment in the power system, this will cause the power to deviate from the initial state[2].

1.3 Calculation and analysis of power and power factor

(1) Single-phase circuit

In a single-phase circuit, the active power P is also generally called average power, actual power or real power, which indicates the effective power of the load during actual operation. Its expression:

$$P=\frac{1}{T}\int_{t_0}^{T+t_0} p(t)dt=\frac{1}{T}\int_{t_0}^{T+t_0} u(t)i(t)dt \quad (3)$$

Where $u(t)$ 、 $i(t)$ are the instantaneous values of the voltage and current at the time t of the phase circuit, respectively, so $p(t)$ is the instantaneous power of the phase circuit at time t. Similarly, the process of discretizing its sampling point to N is as follows:

$$P=\frac{1}{T}\int_{t_0}^{T+t_0} u(t)i(t)dt=\frac{\Delta t}{T}\sum_{k=1}^{N}u_k i_k=\frac{1}{N}\sum_{k=1}^{N}u_k i_k \quad (4)$$

For a single-phase sinusoidal circuit, the active power can be defined as:

$$P=UI\cos\varphi=S\cos\varphi \quad (5)$$

In the above formula, U and I are respectively the voltage rms value and the current rms value of the phase circuit, and the product of U and I is its apparent power S, and φ is the phase angle of the current lag voltage in the phase circuit, that is, the power factor angle.It is easy to derive the calculation formula of the power factor $\cos\varphi$ of the phase circuit as

$$\cos\varphi=\frac{P}{S} \quad (6)$$

Then the calculation formula of the reactive power Q of the single-phase sinusoidal circuit is:

$$Q=UI\sin\varphi \quad (7)$$

(2) Three-phase circuit

In a three-phase symmetrical circuit, the three-phase total active power can be calculated by multiplying the active power of any one of the three phases by three. In the actual three-phase circuit, since the load connected by the three phases is not necessarily symmetrical, the system will calculate the three-phase total active power according to formula (8).That is, calculate the active power of the three phases and add them separately.

$$P_{\text{total}}=P_A+P_B+P_C=\frac{1}{N}\left(\sum_{k=1}^{N}u_{kA}i_{kA}+\sum_{k=1}^{N}u_{kB}i_{kB}+\sum_{k=1}^{N}u_{kC}i_{kC}\right) \quad (8)$$

Similarly, the total apparent power of the three-phase circuit is the sum of the three-phase apparent power of A、B、C, as shown below:

$$S_{\text{total}}=U_A I_A+U_B I_B+U_C I_C \quad (9)$$

By combining equations (8) and (9), the total power factor $\cos\varphi$ of the three-phase circuit can be derived as:

$$\cos\alpha=\frac{P_{\text{total}}}{S_{\text{total}}} \quad (10)$$

1.4 Calculation and Analysis of Voltage Qualification Rate and Power Supply Reliability Rate

For urban areas: the reliability of power supply is not less than 99.90%, and the pass rate of residential customers is not less than 96%. For rural areas: the reliability of power supply is not less than 99.589%, and the pass rate of residential customers is not lower than 93.89%.

(1) The voltage pass rate refers to the percentage of the total time of the monitoring point voltage within the qualified range and the total time of the monthly voltage monitoring within one month of

IOP Conf. Series: Earth and Environmental Science **356** (2019) 012005 doi:10.1088/1755-1315/356/1/012005

the grid operation. The voltage monitoring point should select a representative substation 10kV bus and user as the voltage quality assessment point. For low voltage (380/220) users, the monitoring point should be set at the head end of the representative low voltage line and end. The calculation formula is as shown in equation (11)[3].

voltage pass rate(%)=voltage qualified time/assessment time × 100% (11)

(2) During the statistical period, the power supply reliability rate (RS-3) does not account for the need to limit the power due to insufficient system power. The calculation formula is as shown in equation (11)[4].

power supply reliability(%)=(1-power outage time/statistics time)×100% (12)

2. System software and hardware design

A virtual instrument is an instrument that is tightly integrated with a computer. Virtual instruments offer greater scalability and flexibility than traditional instruments.As a standard data acquisition and instrument control software, LabVIEW is widely accepted by academia, industry and research laboratories[5].

In the virtual instrument technology, the main function of the hardware module is to convert various analog quantities (current, current, liquid level, water temperature, pressure and other physical quantities) and some digital quantities input by the outside world into a suitable hardware system through a certain hardware system. The main hardware modules of the power monitoring and management system designed in this paper include: three-phase voltage sensor, three-phase current sensor, data acquisition card and computer.

Due to the large electric load of the residential area, air conditioning in the residential area is a high-power load, especially in the summer, air conditioning is the main cause of overloading of the power system.Therefore, the load selected in this experiment is Gree's split cold air type floor air conditioner(KF-120LW/E(12368L)A1-N2).The performance indicators of air conditioners are shown in Table 1.

Table 1 .Performance indicators of air conditioning

index	parameter
Cooling capacity	1200W
Rated voltage	380V 3N~
Rated frequency	50Hz
cooling rated frequency	336W
exhaust side maximum working pressure	2.5MPa
suction side maximum working pressure	0.6MPa

The acquisition circuit consists of three voltage transformers, three current transformers and a data acquisition card. The specific experimental wiring is shown in Figure 1.

Figure 1. Acquisition circuit

3. Power Monitoring and Management Platform

3.1 Data acquisition card collection and analysis

Data collection is collected by the acquisition card, and the collected data is written into the queue. This queue uses the cyclic producer consumer model[6].This block diagram is the producer. In this data block diagram, the data error cluster can be transmitted and displayed, and the text problem is displayed. When the loop stops collecting, the data queue is released.The block diagram of the data acquisition program is shown in Figure 2.

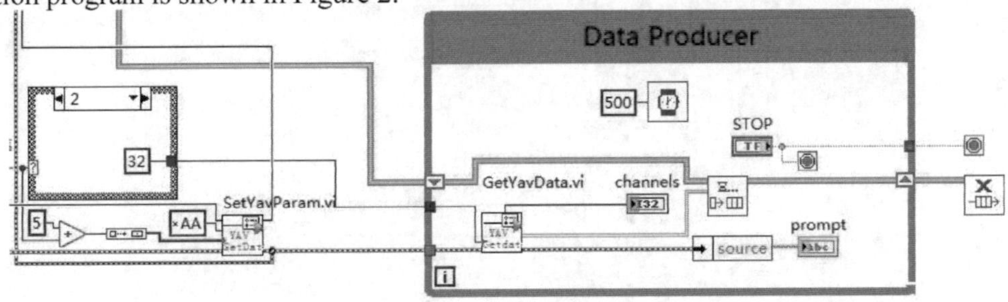

Figure 2. Data acquisition block diagram

The block diagram is based on the data information generated by the producer, and the data is parsed for analysis. This cycle is the consumer of the previous data producer. The block diagram is the producer of the next producer consumer, and the data is passed to the next. The consumer calculates the average value of the collected data, then performs a proportional operation and multiplies the corresponding scale factor to obtain the corresponding data information value. The data analysis program block diagram is shown in Figure 3.

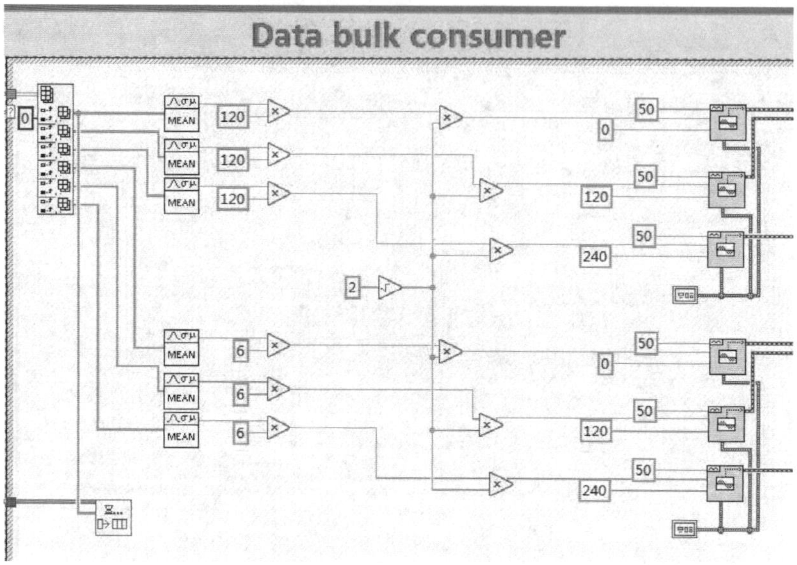

Figure 3. Data analysis block diagram

3.2 real-time acquisition interface

The main acquisition parameters of the real-time acquisition interface include three-phase voltage values and three-phase current values.Therefore, when designing the front panel, you need to select the "meter" in "new" in the "Control" panel, which are named "A phase voltage effective value","B phase voltage effective value","C phase voltage effective value"," A phase current effective value","B phase current effective value","C phase current effective value".The front panel of the real-time acquisition interface is shown in Figure 4.

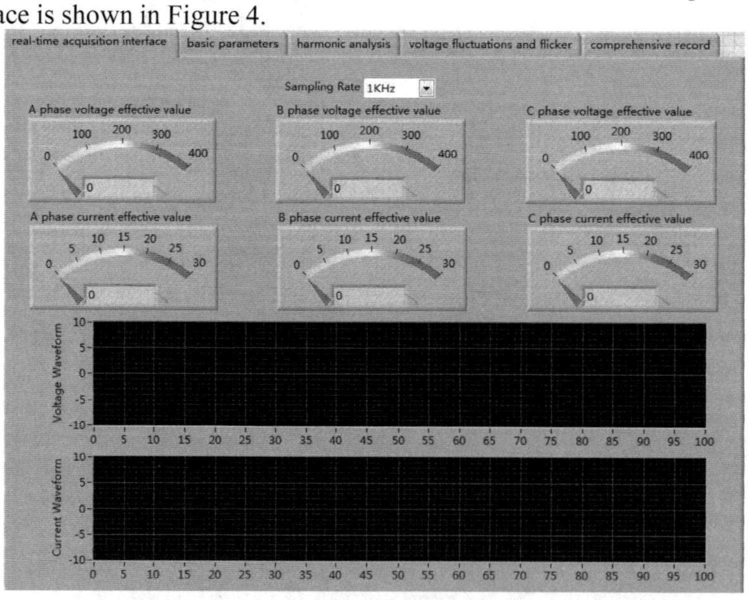

Figure 4. Real-time acquisition interface front panel

The basic parameters include voltage and current, frequency, power, electrical, and imbalance analysis for each phase. Therefore, when designing the front panel, you need to select the "numeric display control" in "New" in the "Control" panel, arrange these functions according to the corresponding needs, and select "New" in the "Control" panel. Therefore, the design of the front panel of the basic parameter module is completed. The front panel of the basic parameter module is shown in Figure 5.

IOP Conf. Series: Earth and Environmental Science **356** (2019) 012005 doi:10.1088/1755-1315/356/1/012005

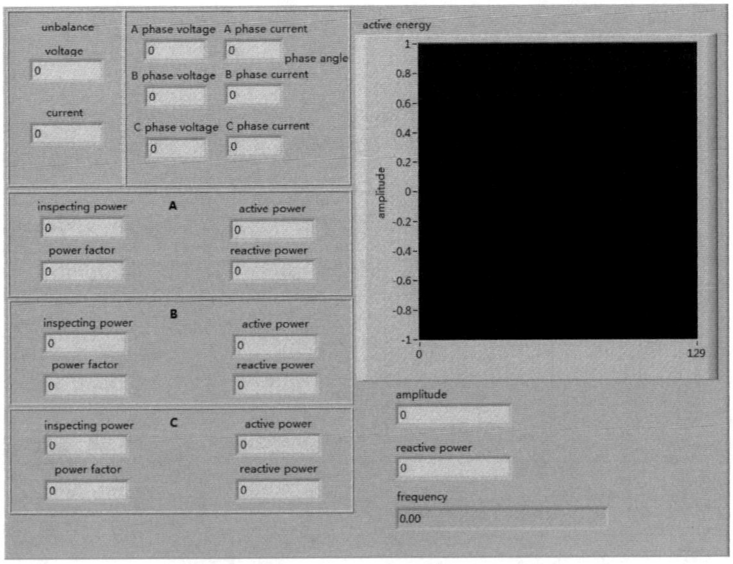

Figure 5. Basic parameters front panel

3.3 Comprehensive record

The comprehensive record include the storage of measurement data, the recording of the power-off time, the reliability of the power supply, the time of the deadline, the phase loss time, the pass rate, and the hourly record.When designing the front panel, you need to select the “Numerical Display Controls” in the “Controls” panel, which are named as the upper limit time, the lower limit time, the phase loss time, the total time, the pass rate, the power outage time and the power supply reliability[7]. The front panel of the comprehensive record is shown in Figure 6.

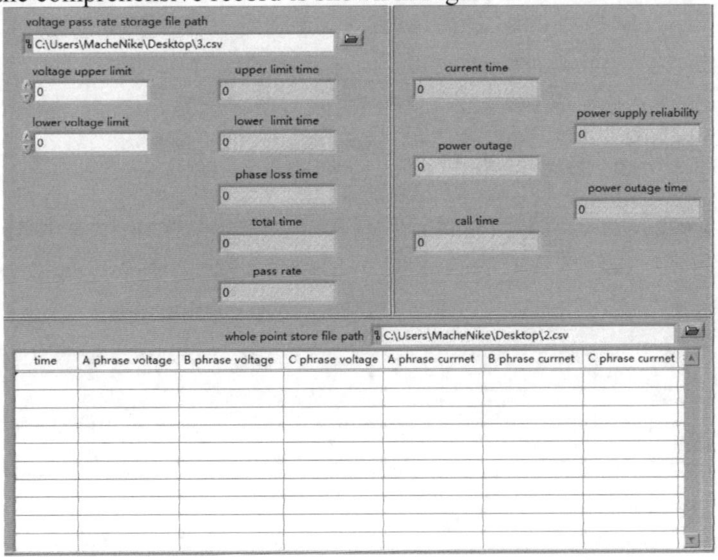

Figure 6. Comprehensive record front panel

The system will record the total time of the collection in real time, using the time format string, and then processing the time and date string, and extracting the day string to judge. When the current date is different, the data information will be stored. The storage block diagram is shown in Figure 7.

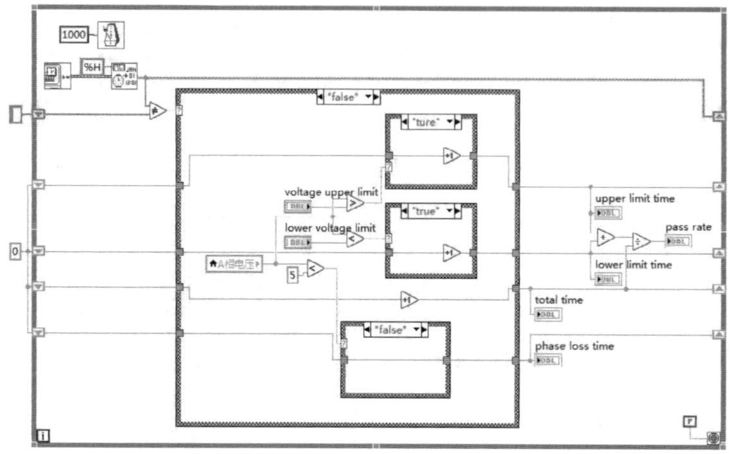

Figure7. The storage block diagram

4 .Conclusion

This paper first introduces the measurement and calculation methods of various indicators of power monitoring and management, then describes the hardware design of the system and the characteristics of the LabVIEW software platform, and uses LabVIEW to write the program for communication with the data acquisition card, and then writes the power monitoring and management system. Each parameter measurement module and management module includes real-time acquisition interface module, basic parameter module and integrated recording module. Finally, each sub-module is connected together, and the design of the community electric energy monitoring and management system is completed through the acquisition circuit.

5. References

[1] Wang M, Sun Y A practical, precise method for frequency tracking and phasor estimation[J]. Power Delivery, IEEE Transactions on, 2004, 19(4):1547-1552.

[2] Liu Xingyuan. Factors affecting power quality and improvement measures [J]. Science and Technology Information, 2013 (15): 401 + 466.

[3] Liu Tingting. Design and implementation of voltage monitoring and integrated management system of Meishan Company [D]. University of Electronic Science and Technology, 2013.

[4] Wang Ruiping,Liu Xiaohui,Zhang Zhong,Qian Zhenyu.Application of Voltage and Current Real-time Monitoring Device in Railway Battlefield Power Supply[J].Electric Railway,2017,28(06):13-15.

[5] Elliott C,Vijayakumar V, Zink W, et al. National Instruments LabVIEW: A Programming Environment for Laboratory Automation and Measurement[J]. Journal of the Association for Laboratory Automation, 2007, 12(1):17-24.

[6] Wu Ruidong. Distributed optical fiber vibration detection and positioning system based on linear Sagnac [D]. Taiyuan University of Technology, 2017.

[7] Li Guodong. Research on Modern Power Quality Comprehensive Evaluation Method [D]. North China Electric Power University (Beijing), 2010.

Power Management System Based on Virtual Power Plant

Xiaofeng Chen[1],Guanlu Yang[2], Yajing Lv[3] and Zehong Huang[4]

College of Information and Science Engineering,Huaqiao University, Xiamen , China
E-mail:410417815@qq.com

Abstract. In recent years, the temperature-controlled load capacity has increased rapidly in China, and it has also brought huge regulatory potential. In order to effectively dispatch and manage a large number of demand response resources, this paper studies the energy efficiency management system based on virtual power plant based on demand response technology and virtual power plant theory, and effectively uses the management system to temperature. Collecting and controlling the power parameters of the load-control resources, making full use of the potential for reducing the demand side load, to achieve the "shift peak filling" of electricity, which is conducive to alleviating the contradiction between power supply and demand and ensuring the safe and stable operation of the power grid.

1. Introduction

In the context of power reform, a "virtual power plant(VPP) " technology introduced to accelerate the development and utilization of new energy sources and effective grid-connected purposes, which aggregates and coordinates the management of distributed energy resources（DERs） unitswithin the jurisdiction, using DERs Fast and flexible adjustment features and complementary functions can improve the economic and security of the power system [1-2].The essence of VPP is to combine advanced advanced management, control, communication and information technologies with traditional power system technologies to improve the safety, reliability, economy and stable operation of the power grid.

With the advancement of power system marketization reform and the adjustment of national energy policy, based on smart grid technology support, cooperation and information exchange between power network and power market, power network and power users are getting closer and closer [3].Demand Side Management (DSM) is an effective method and measure to reduce power consumption and reduce voltage shortage in the same power consumption scenario, thereby ensuring balance between supply and demand and reducing power and electricity costs.

In order to effectively dispatch and manage a large number of demand response resources, this paper studies the energy efficiency management system based on VPP based on demand response technology and VPP theory, and effectively utilizes the management system for temperature control load. Energy parameter collection and control of resources to achieve energy efficiency management of demand side loads, and thereby further enhance the power system's ability to adjust.

2. Overall design

This scheme designs a three-phase three-phase electric load energy parameter acquisition system, including voltage, current, various types of electric power, active power, reactive power, power factor, voltage imbalance, current imbalance, zero sequence Current, power outage time, and the maximum value of voltage and current in 15 minutes and the moment of occurrence.In terms of control, the 3-way load is switched and lifted to support independent control. In the wireless transmission function

Content from this work may be used under the terms of the Creative Commons Attribution 3.0 licence. Any further distribution of this work must maintain attribution to the author(s) and the title of the work, journal citation and DOI.
Published under licence by IOP Publishing Ltd

of the data, the measured data can be transmitted to the upper computer through the ZigBee wireless module, and the command control load can be received from the upper computer.The overall design framework is shown in Figure 1.

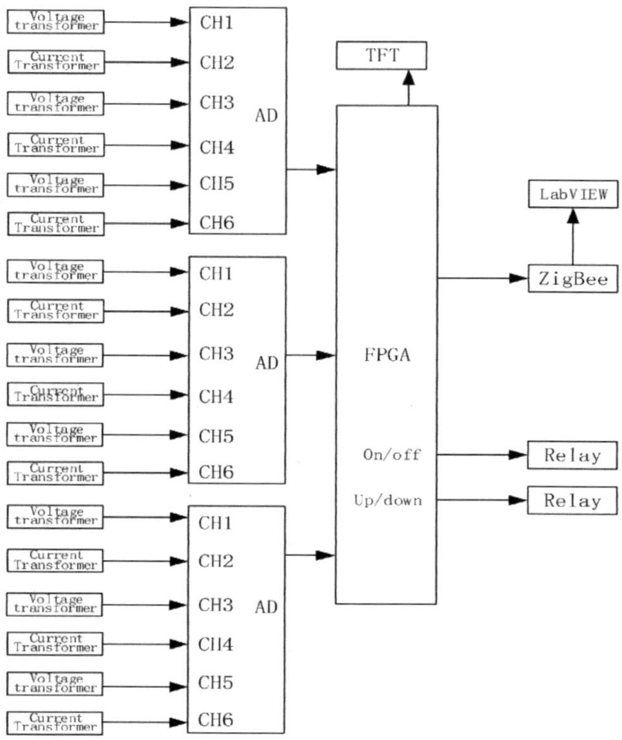

Figure 1. Overall design framework

3. NIOS module design

According to the design requirements, the components to be added from the SOPC Builder are mainly included: CLK, CPU, SDRAM memory module, EPCS storage, ID number, button KEY, UART interface, LCD display, control signal, etc. After adding the components required by the system, assign the base address of each component of the Nios II system.The module configuration and its base address are shown in Figure 2.

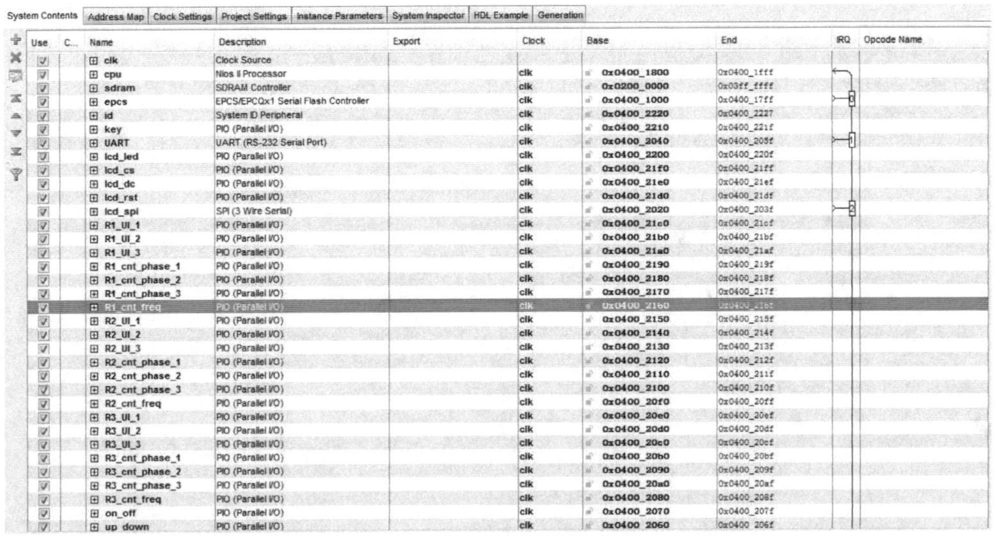

System Contents | Address Map | Clock Settings | Project Settings | Instance Parameters | System Inspector | HDL Example | Generation

Use	C...	Name	Description	Export	Clock	Base	End	IRQ	Opcode Name
☑		clk	Clock Source						
☑		cpu	Nios II Processor		clk	0x0400_1800	0x0400_1fff		
☑		sdram	SDRAM Controller		clk	0x0200_0000	0x03ff_ffff		
☑		epcs	EPCS/EPCQx1 Serial Flash Controller		clk	0x0400_1000	0x0400_17ff	0	
☑		id	System ID Peripheral		clk	0x0400_2220	0x0400_2227		
☑		key	PIO (Parallel I/O)		clk	0x0400_2210	0x0400_221f		
☑		UART	UART (RS-232 Serial Port)		clk	0x0400_2040	0x0400_205f	1	
☑		lcd_led	PIO (Parallel I/O)		clk	0x0400_2200	0x0400_220f		
☑		lcd_cs	PIO (Parallel I/O)		clk	0x0400_21f0	0x0400_21ff		
☑		lcd_dc	PIO (Parallel I/O)		clk	0x0400_21e0	0x0400_21ef		
☑		lcd_rst	PIO (Parallel I/O)		clk	0x0400_21d0	0x0400_21df		
☑		lcd_spi	SPI (3 Wire Serial)		clk	0x0400_2020	0x0400_203f	2	
☑		R1_UI_1	PIO (Parallel I/O)		clk	0x0400_21c0	0x0400_21cf		
☑		R1_UI_2	PIO (Parallel I/O)		clk	0x0400_21b0	0x0400_21bf		
☑		R1_UI_3	PIO (Parallel I/O)		clk	0x0400_21a0	0x0400_21af		
☑		R1_cnt_phase_1	PIO (Parallel I/O)		clk	0x0400_2190	0x0400_219f		
☑		R1_cnt_phase_2	PIO (Parallel I/O)		clk	0x0400_2180	0x0400_218f		
☑		R1_cnt_phase_3	PIO (Parallel I/O)		clk	0x0400_2170	0x0400_217f		
☑		R1_cnt_freq	PIO (Parallel I/O)		clk	0x0400_2160	0x0400_216f		
☑		R2_UI_1	PIO (Parallel I/O)		clk	0x0400_2150	0x0400_215f		
☑		R2_UI_2	PIO (Parallel I/O)		clk	0x0400_2140	0x0400_214f		
☑		R2_UI_3	PIO (Parallel I/O)		clk	0x0400_2130	0x0400_213f		
☑		R2_cnt_phase_1	PIO (Parallel I/O)		clk	0x0400_2120	0x0400_212f		
☑		R2_cnt_phase_2	PIO (Parallel I/O)		clk	0x0400_2110	0x0400_211f		
☑		R2_cnt_phase_3	PIO (Parallel I/O)		clk	0x0400_2100	0x0400_210f		
☑		R2_cnt_freq	PIO (Parallel I/O)		clk	0x0400_20f0	0x0400_20ff		
☑		R3_UI_1	PIO (Parallel I/O)		clk	0x0400_20e0	0x0400_20ef		
☑		R3_UI_2	PIO (Parallel I/O)		clk	0x0400_20d0	0x0400_20df		
☑		R3_UI_3	PIO (Parallel I/O)		clk	0x0400_20c0	0x0400_20cf		
☑		R3_cnt_phase_1	PIO (Parallel I/O)		clk	0x0400_20b0	0x0400_20bf		
☑		R3_cnt_phase_2	PIO (Parallel I/O)		clk	0x0400_2090	0x0400_209f		
☑		R3_cnt_phase_3	PIO (Parallel I/O)		clk	0x0400_20a0	0x0400_20af		
☑		R3_cnt_freq	PIO (Parallel I/O)		clk	0x0400_2080	0x0400_208f		
☑		on_off	PIO (Parallel I/O)		clk	0x0400_2070	0x0400_207f		
☑		up_down	PIO (Parallel I/O)		clk	0x0400_2060	0x0400_206f		

Figure 2.module configuration and base address map

In Figure 2, CLK is the clock module, CPU is the system control unit module, SDRAM stores the running program, EPCS stores the program after curing, ID is the ID number of the CPU, KEY is the button module, UART is the serial port module, and data is carried out with ZIGBEE The transmission and interaction, the LCD display is used to control the screen display, and the R series pins collect 3 channels of measured data, including voltage, current, phase, frequency, and so on. On_off and on_down are control signals for switching and lifting.

（1）RMS measurement

The rms measurement includes voltage rms measurement and current rms measurement. The rms measurement mainly applies the square and re-opening operation. In this design, the RTL function simulation of the squared summation of the rms measurement is shown in Figure 3.

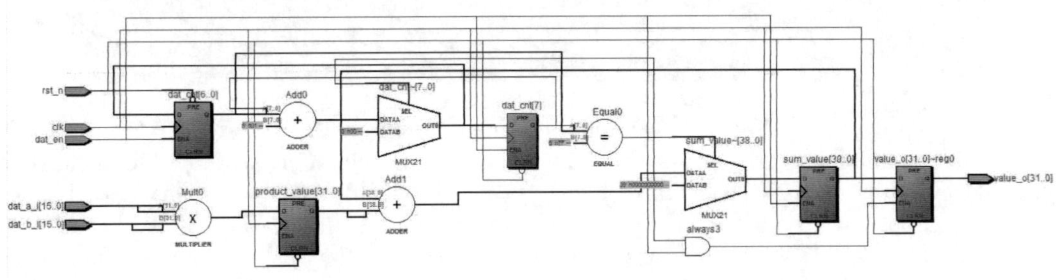

Figure3.Functional simulation of the sum of squares

In Figure 3, rst_n is the reset signal, clk provides the clock signal to the module, and dat_en is the enable signal. 16-bit dat data signal input, Mult0 performs the product of the dat input signal, product_value completes the buffering of the product result. With the participation of the dat_en enable signal, the dat_cnt register controls the number of superpositions, and the dat_cnt register and the enable signal dat_en jointly control the product_value. The accumulation operation, the sum of the operations is stored in the sum_value register. When the dat_cnt counts to a certain value, the value_o register latches the value of the sum_value register as the squared data output.

（2）Frequency measurement

There are two main methods for measuring the frequency. This paper mainly uses the period measurement method. In the design, the sig_reg register and the sig_reg_1 register respectively collect the rising edge of the clock input signal, and the detection of the rising edge is completed by the pos_sig gate circuit, and the clock sys_clk is counted between the two rising edges, and the obtained technical result and sys_clk are obtained. The period is multiplied to measure the period value of the signal, and the reciprocal is the frequency. The Add0 adder and ref_cnt implement the superposition of the rising edge data, and the cnt_sys outputs the measurement result. The functional simulation RTL diagram of frequency measurement is shown in Figure 4:

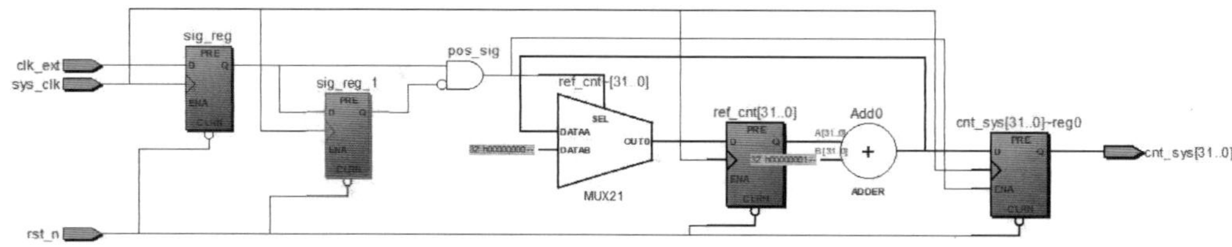

Figure4. Functional simulation RTL diagram of frequency measurement

4. Simulation analysis

This design uses Modelsim10.1c to perform functional simulation verification and functional simulation of AD module sampling. The AD sampling simulation diagram is shown in Figure 5.

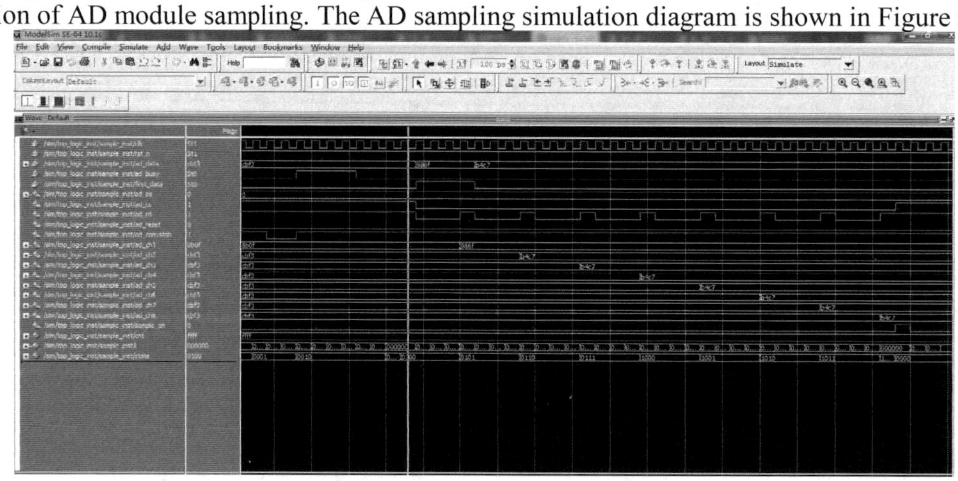

Figure5.AD sampling simulation diagram

In the simulation diagram, ad_convsta and ad_convsta share the convst signal to achieve data synchronization sampling of 8 channels. Ad_busy is a busy signal, cs/rd is a read/write signal, and data:db is a data bus. When the ad_convstab signal is a rising edge, the ad_busy signal becomes high. When both the ad_cs bit and the ad_rd signal become low, the first_data signal is simultaneously asserted high, and the chip will read the eight channels of the converted data in turn.

5. Design of power parameter calculation module

Combined with the electrical parameters measured by this design, this paper will introduce the calculation methods of voltage RMS, current RMS, active power, reactive power, apparent power, power factor, frequency and three-phase unbalance.

According to the definition of the effective value of the periodic function, for any one of the periodic functions $x(t)$, the effective value is usually defined by the root mean square value of the $x(t)$ period, as follows:

$$M_{rms} = \sqrt{\frac{1}{T}\int_{t_0}^{t_0+T} x^2(t)dt} \quad (5.1)$$

According to the above formula, the effective values of voltage $u(t)$ and current $i(t)$ of the periodic function can be defined as:

$$I_{rms} = \sqrt{\frac{1}{T}\int_{t_0}^{t_0+T} i^2(t)dt} \tag{5.2}$$

$$U_{rms} = \sqrt{\frac{1}{T}\int_{t_0}^{t_0+T} u^2(t)dt} \tag{5.3}$$

In the formula, T is the period of the voltage and current signal. In order to make the calculation applicable to the computer processing system, the above formula needs to be discretized, and the result is:

$$U_{rms} = \sqrt{\frac{1}{N}\sum_{m=0}^{N-1} u^2(m)} \tag{5.4}$$

$$I_{rms} = \sqrt{\frac{1}{N}\sum_{m=0}^{N-1} i^2(m)} \tag{5.5}$$

The meaning of N in the formula is the total number of samples after a period of dispersion, and $u(m)$ and $i(m)$ are the sampled values of the voltage signal and the current signal at time m, respectively. After discrete, the effective values of voltage and current can be calculated simply by computer control system.

In an AC circuit, the apparent power does not represent the power actually consumed by the AC circuit, but only the maximum power that the circuit may provide or the maximum active power that the circuit may consume.We refer to the product of the rms voltage and the rms current in the AC circuit as the apparent power. The formula is:

$$S = U_{rms} I_{rms} \tag{5.6}$$

In an AC circuit, active power is also referred to as average power, meaning the average of the integrals of instantaneous power consumed or emitted over a period.Record the active power as P, the formula is

$$P = S\cos\varphi \tag{5.7}$$

Reactive power is relatively abstract, and we believe that reactive power is used in electric and magnetic fields in circuits and is used to establish and maintain electrical power in electrical equipment [4].

$$Q = S\sin\varphi \tag{5.8}$$

In the actual situation, the load connected to the three-phase circuit is not necessarily symmetrical, resulting in the asymmetry of the three-phase circuit. Therefore, the calculation method of the three-phase total active power of the present design is to add the active power of each phase to obtain the true three. The total active power of the phase.

$$P_{\text{total}} = P_A + P_B + P_C \tag{5.9}$$

Similarly, the calculation method of total reactive power and total apparent power can be obtained.

$$Q_{\text{total}} = Q_A + Q_B + Q_C \tag{5.10}$$

$$S_{\text{total}} = S_A + S_B + S_C \tag{5.11}$$

The three-phase unbalance degree refers to the degree of three-phase unbalance in the three-phase shop system, and is expressed by the root mean square percentage of the negative-sequence fundamental wave component or the zero-sequence fundamental wave component of the voltage and current and the positive-sequence fundamental wave component[5]. Generally, we can roughly judge the operating condition of the three-phase circuit according to the magnitude of the three-phase unbalance rate. If the unbalance rate is lower than 2%, it is generally caused by the single-phase load imbalance of the three-phase circuit[6-7]. The calculation formula for the three-phase voltage and current imbalance is as follows:

$$\beta_{\mathrm{u}} = \frac{U_2}{U_1} \times 100\% \tag{5.12}$$

$$\beta_{\mathrm{I}} = \frac{I_2}{I_1} \times 100\% \tag{5.13}$$

Since the input power of the data center adopts a three-phase four-wire system, when the three-phase system is unbalanced, the zero-line current will flow through the zero-sequence current, that is, there is a zero-sequence component in the three-phase system. Therefore, when calculating the imbalance of the three-phase four-wire system, the symmetrical component method must be used to determine the components of the positive and negative sequence of the fundamental voltage and the fundamental current, respectively, and then according to the formula (5.12), (5.13) Find the imbalance of the three-phase voltage and current.The formula for calculating the positive sequence component and the negative sequence component is as follows:

$$\dot{U}_1 = \frac{1}{3}\left(\dot{U}_A + \alpha \dot{U}_B + \alpha^2 \dot{U}_C\right) \tag{5.14}$$

$$\dot{U}_2 = \frac{1}{3}\left(\dot{U}_A + \alpha^2 \dot{U}_B + \alpha \dot{U}_C\right) \tag{5.15}$$

$$\dot{I}_1 = \frac{1}{3}\left(\dot{I}_A + \alpha \dot{I}_B + \alpha^2 \dot{I}_C\right) \tag{5.16}$$

$$\dot{I}_2 = \frac{1}{3}\left(\dot{I}_A + \alpha^2 \dot{I}_B + \alpha \dot{I}_C\right) \tag{5.17}$$

In the above formula, α is the twiddle factor, $\alpha = -\frac{1}{2} + \mathrm{j}\frac{\sqrt{3}}{2}$, $\alpha^2 = -\frac{1}{2} - \mathrm{j}\frac{\sqrt{3}}{2}$, $\dot{U}_1$ and $\dot{I}_1$ are the positive sequence components of voltage and current, respectively, $\dot{U}_2$ and $\dot{I}_2$ are the negative sequence components of voltage and current, respectively, and $\dot{U}_A$, $\dot{U}_B$ and $\dot{U}_C$ are the fundamental components of the three-phase voltages of A, B and C, respectively. $\dot{I}_A$, $\dot{I}_B$, and $\dot{I}_C$ are the fundamental components of the three-phase currents of A, B, and C, respectively.

There are two main methods for calculating the zero-sequence current detection, namely zero-sequence current transformer detection method and three-phase current synthesis method[8].Both methods have their own advantages. In this paper, the three-phase current synthesis method is used to calculate the zero-sequence current according to the current conditions, and the calculation formula can be obtained as

$$I = \frac{I_a}{k} + \frac{I_b}{k} + \frac{I_c}{k} \tag{5.18}$$

I is the zero sequence current, I_a is the phase A voltage, I_b is the phase B voltage, I_c is the phase C voltage, and k is the ratio of the three, three, and four phase current transformers.

6. Conclusions

In the traditional power system, energy flows from the power generation side to the user side only, and the emergence of demand response technology enables the demand side resources to participate in the power market, that is, to achieve energy interaction, reduce power generation costs, and make power

supply more efficient. clean. At the same time, the concept of VPP brings together a variety of distributed power sources, controllable loads and energy storage units to participate in the dispatching and management of the electricity market and grid operation as another special power plant.
Based on the integration of temperature control load, demand response and VPP concept, based on demand response technology and VPP theory, the temperature control load is taken as the research object, and the energy efficiency management system based on VPP is built, which effectively utilizes the management system to load resources. The power parameter acquisition and control is used to fully utilize the potential side load reduction potential, thereby further enhancing the power system's regulation capability.

7. References

[1] Yu Shuang,Wei Zhinong, Sun Guoqiang, et al. A Bidding Model for a Virtual Power Plant Considering Uncertainties[J]. Automation of Electric Power Systems,2014,38(22):43-49.
[2] PANDZIC H,KUZLE I,CAPUDER T. Virtual power plant mid-term dispatch optimization[J]. Applied Energy,2013,101:134-141.
[3] Zhang Qin, Wang Xifan, fu Min, et al. Smart Grid from the Perspective of Demand Response[J]. Automation of Electric Power Systems.2009,33(17):49-55.
[4] Bu Weizhong. Reactive power compensation and power factor improvement[J].China High-Tech Enterprises., 2009, 1(07):110-111.
[5] National Standardization Management Committee. GB/T 15543-2008,Power quality t hree-phase voltage imbalance[S].Beijing: China Standards Press,2008.
[6] Jing Yong,Liang Chen,Shuangyan Chen.Modeling of Home Appliance for Power Distribution System Harmonic Analysis [M].IEEE Trans.Power Delivery, vol. 25, pp. 3147~3115, 2010
[7] Roger C.Dugan, Mark F. McGranaghan， et.al.Power system power quality[M].Ouyang Sen, Translation. Beijing: China Electric Power Press，2013.8
[8] Feng Xiaoguang. On the Protection of Zero Sequence Current[J].Journal of Library and Information Science, 2009, 19(27):210-211

Lanthanum leachate concentration from spent catalysts

Caterine Donoso[1], Alex Maza[2] and Ricardo Urrutia-Goyes[3]

Departamento de Ciencias de la Energía y Mecánica, Universidad de las Fuerzas Armadas ESPE, Av. Gral. Rumiñahui s/n, P.O. Box 171-5-231B, Sangolqui 171103, Ecuador

E-mail: erurrutia@espe.edu.ec (*R.U.-G.*)

Abstract. Spent catalysts generate leachates that cause very serious environmental problems. This study shows the evaluation of the influence of applying a base to the leach recovered from an exhausted catalyst obtained from the "Esmeraldas State Refinery" in Ecuador. The main goal was to increase the percent concentration by mass of lanthanum and in turn reduce the concentration of aluminum through its precipitation. The agent used was 4.0 and 5.0 M sodium hydroxide and the pH was adjusted to stimulate the precipitation. Results show that the agent used is a precipitator of aluminum since the concentration of lanthanum increased up to 44%. The use of metal hydroxides in leachates is a valid alternative for the removal of heavy metals in order to prevent the contamination of the environment.

1. Introduction

Different types of catalysts are commonly used in the chemical, petrochemical, and petroleum refining industry. These reagents seek an increased reaction rate and a modified selectivity process in different chemical processes in order to allow an improved energy efficiency and hence reduced costs. Catalysts often need to be replaced, due to deactivation, every two to three years. Globally, approximately $4x10^8$ kg of spent catalyst are produced every year [1].

Spent catalysts are considered one of the largest hazardous waste in the world due to their toxicity. For instance, the deposition of sulfur, carbon, and heavy metals such as lanthanum, nickel, aluminum, cerium, titanium, vanadium, and iron, among others, have been reported [2]. Catalysts found in the hydrotreatment unit from "Esmeraldas State Refinery" in Ecuador contain rare earth elements such as Lanthanum and Cerium. These two elements are used in many industrialized countries in the technology business and hence the treatment of these types of spent catalysts is very important in order to decrease the amount of hazardous waste generated. Other benefits include environmental and economic aspects [3].

The recovery of metals from spent catalysts can be achieved by two main methods i.e. hydrometallurgical and pyrometallurgical. In the former, metals are dissolved by leaching the catalyst using an acid or a base in a way that such metals can be recovered as compounds; in the latter, heat is used as a treatment method, such as calcination or smelting, to separate the metals [4]. In the hydrometallurgical method, sodium hydroxide can be used as the precipitating agent [3] since it yields the reaction of some hazardous metals such as aluminum, silicon, iron, nickel, and copper [5].

The main goals of this research project are: a) to precipitate aluminum in the form of hydroxide, from a leachate obtained from a spent catalyst in the hydrotreatment unit from "Esmeraldas State Refinery" in Ecuador; and as a consequence, b) to increase the concentration in mass of lanthanum in the resulting solution.

Content from this work may be used under the terms of the Creative Commons Attribution 3.0 licence. Any further distribution of this work must maintain attribution to the author(s) and the title of the work, journal citation and DOI.
Published under licence by IOP Publishing Ltd

IOP Conf. Series: Earth and Environmental Science **356** (2019) 012007 doi:10.1088/1755-1315/356/1/012007

2. Methodology

Leachate samples were obtained from a spent catalyst in the hydrotreatment unit from "Esmeraldas State Refinery" in Ecuador. The purpose of the recovery of lanthanum is the subsequent impregnation of the metal in alumina supports. In previous studies [6], the spent catalyst was treated with nitric acid at a concentration of 3.0 M, temperature of 20°C, with a solids content of 20% and a leaching time of one hour at 300 rpm; to these conditions the recovery percentage of lanthanum was 99.44%. The analytical technique used for the determination of lanthanum both in the extract as well as in the solid fraction was ICP-OES. The concentration of lanthanum in three extracts leached with nitric acid and which were the starting solutions for the present investigation are: 2,055 ppm, 6,253 ppm and 6,303 ppm of lanthanum and each of them with a volume of 45 mL.

In order to precipitate aluminum by using sodium hydroxide and distilled water (aiming to remove any aluminum hydroxide residue from the lanthanum-rich solution), two solutions with 4.0 M and 5.0 M were prepared [7]. Both solutions were treated until a pH range of 4.0-5.5 was achieved [5]. Constant agitation was maintained for ten minutes at 370 rpm. After the precipitation process, each of the samples underwent decantation for a period ranging between 1 and 4 hours. Finally, samples were filtered for a period ranging between 2 and 24 hours. Every experiment here described was carried out in room temperature (18-21 °C). The resulting pre-treated leachates were analyzed by ICP-MS analytical technique to determine the final percentage of lanthanum content.

Agent sodium hydroxide (98% purity) was obtained from Merck and distilled water was produced with a Boeco distiller (model Dest 4000). The heating plate (model MSH-420), and pH-meter (model PT-70) were obtained from Boeco. ICP-OES analysis was carried out by a Perkin-Elmer device (model Optima 8000). Software packages HSC Chemistry 6.0 was used to plot Pourbaix diagrams, and Minitab to define statistic parameters and tests.

3. Results and Discussion

The Pourbaix diagram is a graphical representation of the potential as a function of pH for a given metal and its species under specific defined thermodynamic conditions [8]. One of the advantages of using such diagram is the ease it provides for the identification of soluble species, ions, and solids. For instance, the separation of lanthanum and aluminum in a solution is defined by pH. Figure 1 shows the pH range 4.40 – 5.21, where the highest concentration of lanthanum was reached. Aluminum forms a hydroxide $[Al_3(OH)_4]^{+5}$ while lanthanum under the same conditions (pH range 4.40-5.21; temperature 20 °C); is found in soluble form (La^{+3}) [9]. The highest removal percentage for lanthanum was 44.31 at a pH of 4.40.

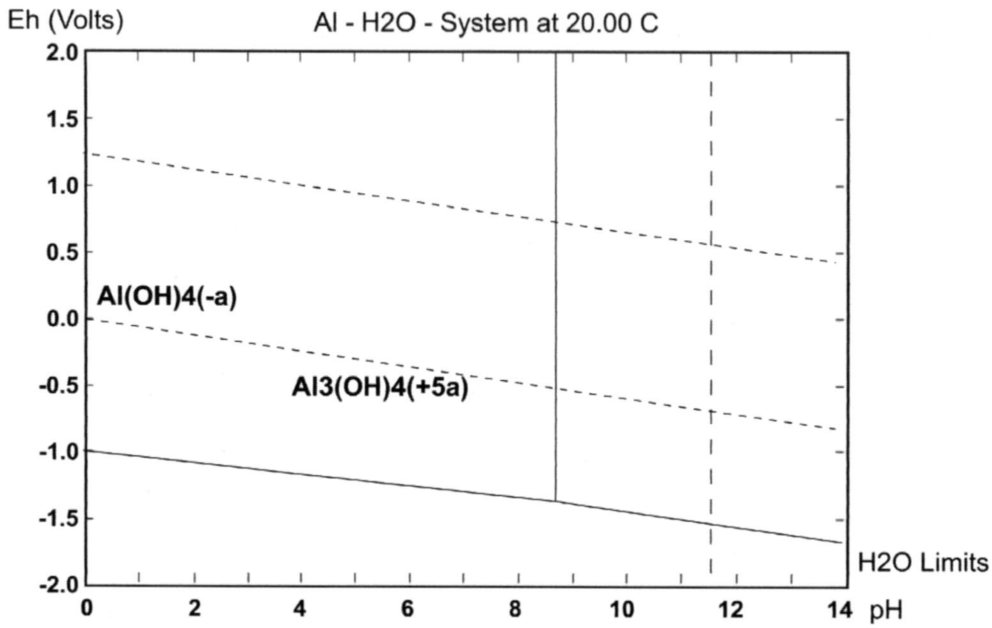

Aluminum

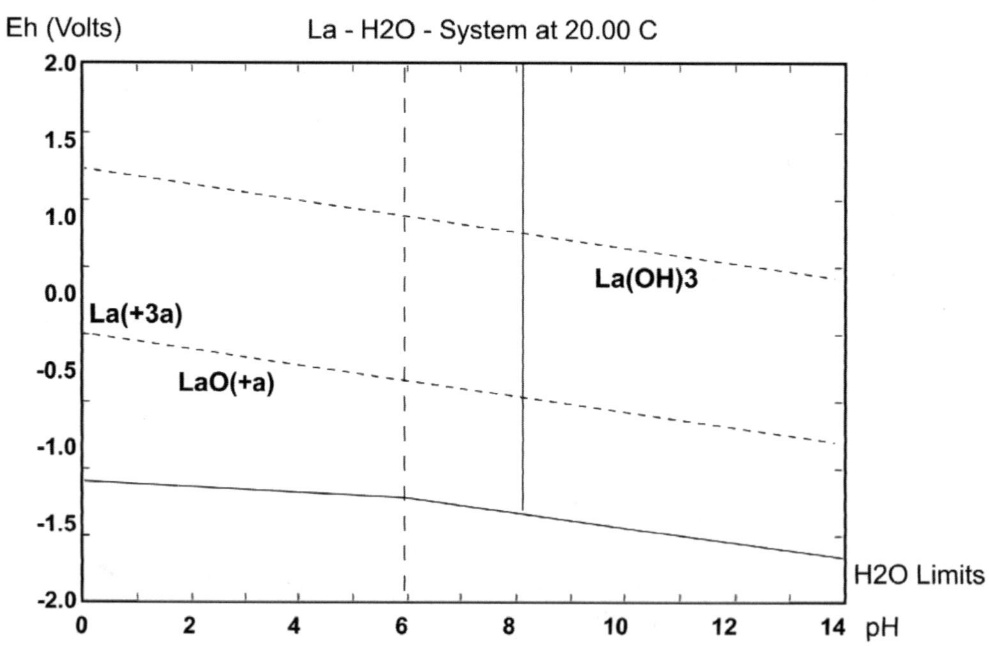

Lanthanum

Figure 1. Purbaix diagrams for aluminum and lanthanum at 20 ℃.

IOP Conf. Series: Earth and Environmental Science **356** (2019) 012007 doi:10.1088/1755-1315/356/1/012007

The optimum pH range reported by Balintova and Petrilakova [5] in order to reach a 98.5% efficiency in the precipitation process of aluminum hydroxide is 4.0-5.5. The described experimentation has allowed pH values close to the ones aforementioned by titration of the studied leachates with NaOH at concentrations of 4.0 and 5.0 M. Moreover, Figure 2 shows that the percentage of removal of lanthanum (from solid phase towards solution) decreases as pH values gradually increase. This means that, at pH values greater than 5.5, the concentration of lanthanum in the solution decreases, favoring its precipitation. According to previous studies carried out by Zhang and Honaker [9], for the same pH range (4.0-5.5), the percentage of removal of lanthanum (from solution towards solid phase) increases. This means that the precipitation of lanthanum is promoted at pH values above 5.5, which is consistent with the results obtained in this experimentation.
Finally, it was found that once a pH value close to 4.0 was reached by treating samples with NaOH at 5.0 M as precipitator agent, the addition of a volume greater than 0.03 mL in the titration process can cause a sudden change in pH up to values greater than 6, which causes co-precipitation of lanthanum and aluminum. The percentual removal for lanthanum from the solid phase ($\%R_{La}$) was calculated according to Eq. 1, where w_R, and w_0 represent the removed and initial mass for the metal, respectively.

$$\%R_{La} = \frac{w_R - w_0}{w_R} \times 100 \tag{1}$$

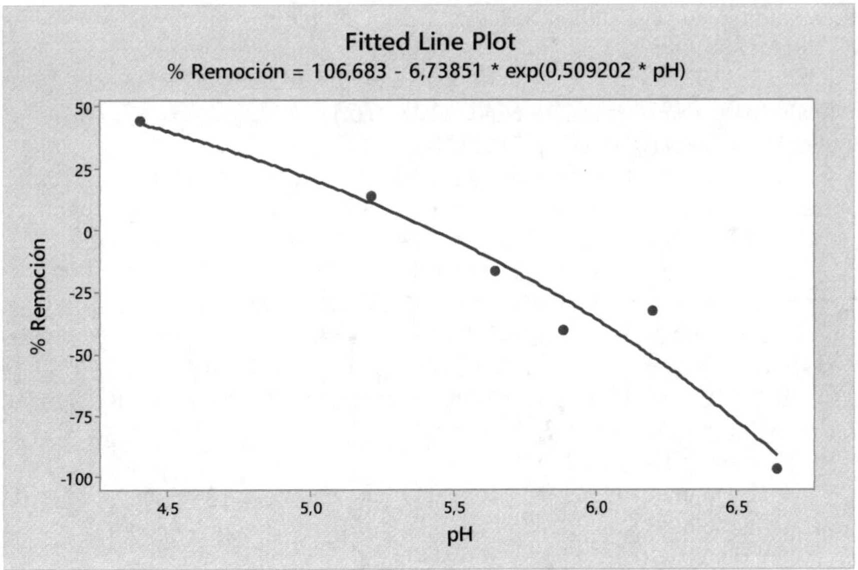

Figure 2. Lanthanum removal in a solution as function of pH.

4. Conclusions

Environmental problems caused by leaching used catalysts can be reduced by treating such wastes with precipitating agents such as hydroxides. The main goal of this research paper was to remove precipitated lanthanum from leachate obtained from spent catalysts. The results showed a removal percentage for lanthanum of up to 44.31 at a pH of 4.40. Additionally, after its evaluation in samples from "Refinería Estatal de Esmeraldas" in Ecuador, the use of sodium hydroxide has proven to be an efficient precipitator for aluminum in solutions obtained from an exhausted catalyst. The aforementioned agent, as well as other metallic hydroxides, must be taken into account in refineries to treat used catalysts and to seek the removal of valuable metals such as lanthanum. The results here presented can be used in the optimization of technologies applied to the selective removal of metals, which have potential economic value.

5. References

[1] Aung, K. M. M., & Ting, Y. P. (2005). Bioleaching of spent fluid catalytic cracking catalyst using Aspergillus niger. *Journal of Biotechnology*, 159–170. https://doi.org/10.1016/j.jbiotec.2004.10.008

[2] IPIECA. (2014). Petroleum refinery waste management and minimization. *Operations Good Practice Serires 2014*, 1–56. Retrieved from www.ipieca.org

[3] Innocenzi, V., Ferella, F., de Michelis, I., & Vegliò, F. (2015). Treatment of fluid catalytic cracking spent catalysts to recover lanthanum and cerium: Comparison between selective precipitation and solvent extraction. *Journal of Industrial and Engineering Chemistry*, 1–6. https://doi.org/10.1016/j.jiec.2014.09.014

[4] Alonso, F., Ramírez, S., Ancheyta, J., & Mavil, M. (2008). Alternativas para la recuperación de metales a partir de catalizadores gastados del hidrotratamiento de hidrocarburos pesados: Un caso de estudio. *Revista Internacional de Contaminacion Ambiental*, *24*(2), 55–69. https://doi.org/0188-4999

[5] Balintova, M., & Petrilakova, A. (2011). Study of pH Influence on Selective Precipitation of Heavy Metals from Acid Mine Drainage. *CHEMICAL ENGINEERING TRANSACTIONS*, *25*, 345–350. https://doi.org/10.3303/CET1125058

[6] Donoso, C., Germania, T., Cárdenas, M. J., Ortiz, V., & Bolívar, A. (2018). Recovery of Heavy Metals from the Spent Catalyst of the Hydrotreating Unit (HDT) for the Use of the Impregnation of Supported Catalysts. *Key Engineering Materials*, *792*, 133–139. https://doi.org/10.4028/www.scientific.net/kem.792.133

[7] Andersson, E., & Hansson, H. (n.d.). Precipitation of reactive aluminium hydroxide from an acidic aluminium sulphate solution by addition of sodium hydroxide. *Symposium a Quarterly Journal in Modern Foreign Literatures*, *6*. https://doi.org/10.1186/gm425

[8] Muñoz, M. (2011). Características y usos de los diagramas de Pourbaix. Retrieved July 4, 2018, from https://riunet.upv.es/bitstream/handle/10251/13708/Características y usos.pdf?sequence=3&isAllowed=y

[9] Zhang, W., & Honaker, R. Q. (2018). Rare earth elements recovery using staged precipitation from a leachate generated from coarse coal refuse. *International Journal of Coal Geology*, *195*, 189–199. https://doi.org/10.1016/j.coal.2018.06.008

Acknowledgments

The authors would like to thank the staff from the laboratory at "Departamento de Metalurgia Extractiva" from Escuela Politécnica Nacional, for their collaboration in the project.

IOP Conf. Series: Earth and Environmental Science **356** (2019) 012008 doi:10.1088/1755-1315/356/1/012008

Simulation analysis of the influencing factors on the wind environment around the hilly terrain

Q Chen[1], Y Liu[1]

1 Wind Engineering Research Center, Xiamen University of Technology, Xiamen, Fujian, 361024, China
Email: 316994464@qq.com

Abstract. The wind environment condition of hilly terrain is closely related to wind resources utilization. In this paper, based on Fluent numerical simulation method, the wind flow characteristics around the hill topography under the influence of varied slopes, incoming wind speeds, heights and roughness are analyzed and summarized. It can be drawn that for the range from the hillside to the top on the windward side, the slope has a significant effect on the wind speed. Within the 100m height above the ground, when the hill is steeper, the wind speed is significantly enhanced. At the top of the hill, the wind speed's maximum increment is 1.5 times of the incoming value. When the wind goes across the hilltop, the factors including the slope, height and roughness, have great influence on the wind environment on the leeward side. When the slope is greater than or equal to 0.5, the phenomenon of eddies present and the range of influence extends downstream. The wind speed weakness trend can also be found when the height and roughness increase.

1. Introduction

Wind in different topographic conditions will produce isthmus effect and slope climbing effect, directly influencing the utilization of dominant wind resources around the terrain or the avoidance of unfavorable wind conditions, i.e. the wind farm microsite or the construction surrounding structures. In China, the hilly terrain account for about 43% of the total land area[1], so the research on the wind environment around topographic affects a very-wide ranging issues.

By means of numerical simulation and wind tunnel test, many scholars simplify the complex terrain into a functional form with variables and then obtain the corresponding parameter data for analysis. First in 1975, Jackson and Hunt[2] studied small curvature hills with uniform rough surface, and obtained the analytical solution of adiabatic turbulent boundary layer of small hills through abstract simplification. They also pointed out that the maximum wind speed point was located at the top of the hills, which provided valuable reference for subsequent studies. In 1989, Gong *etal.*[3] carried out wind tunnel tests on the turbulence of two-dimensional and axial symmetry three-dimensional mountains, which had important reference significance for the study of mountain turbulence. Davenport *etal.*[4] studied the acceleration effect of wind speed on a continuous twelve mountains in two-dimensional condition. Based on the LES model, Uchida and Ohya[5] simulated the wind field in real terrain with RIAM-COMPAC editing program. Kondo *etal.*[6] studied the wind speed increase effect of two-dimensional hills with different slopes in wind tunnel. Yang *etal.*[7] the wind energy resources in complex terrain are evaluated by improving the quasi-static model of Peking University and combining numerical simulation. Chinese architectural structure load standards(GB-50009-2001)[8], the typical mountain topography is mainly considered as ridge and two-dimensional hillside, and the

Content from this work may be used under the terms of the Creative Commons Attribution 3.0 licence. Any further distribution of this work must maintain attribution to the author(s) and the title of the work, journal citation and DOI.
Published under licence by IOP Publishing Ltd

IOP Conf. Series: Earth and Environmental Science **356** (2019) 012008 doi:10.1088/1755-1315/356/1/012008

effect of terrain on the increase of wind speed is taken into account so as to modify the height variation coefficient of wind pressure.

But the researches on wind environmental features around terrain conditions are still lack of systematic and comprehensive study, which needed more comprehensive discussion focusing on the basic terrain types. The method using terrain field testing can only collect limited sample points and the wind tunnel test technique may also use longer time and higher cost, so the numerical simulation has been more widely used due to high efficiency and design parameters. In this study, the computational fluid dynamics software Fluent is used to carry out simulation calculation on wind condition of typical hilly terrain, and factors affecting are comprehensively studied.

2. Calculation model

2.1 Hill terrain

In this paper, the cosine model (see Eq.1) is adopted to simulate the hilly terrain[9], and the mountain model is shown in figure 1. In the figure, *H* is the height of the mountain, *L* is the horizontal distance from the foot of the mountain to the top, $L_1 = L/2$.

$$h = H\cos^2(\pi r/4L_1) \tag{1}$$

2.2 Simulation setup

- Size of the computational domain: 17L(x)×5H(y).
- Boundary conditions:

Inflow boundary condition	Velocity inlet
Outflow boundary condition	Outflow
Surface of the hill model and the bottom	Wall
Surface of the top	Symmetry

- Turbulence model: Standard *k-ε*

The quad-grid is adopted as shown in figure 2.

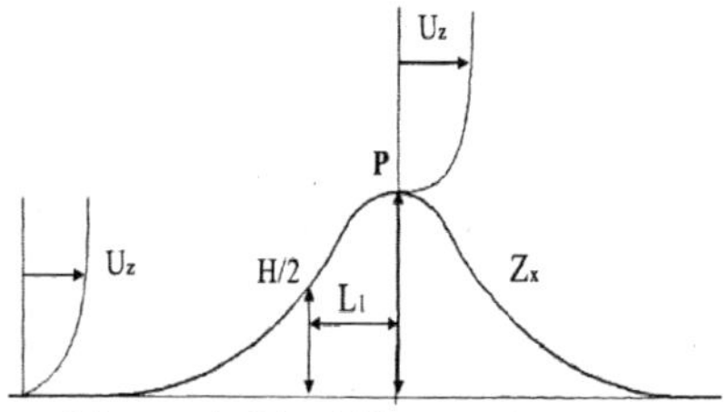

Figure 1. The hilly terrain model

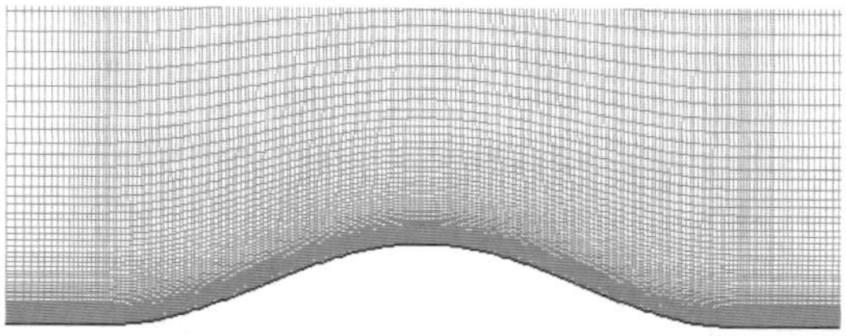

Figure 2. The grids near the hill

2.3 Simulation cases

In this paper, the characteristics of wind environment field around the hilly terrain are discussed based on the influencing factors, such as the slope, wind speed, height and surface roughness. The research conditions are summarized in table 1~4.

Table 1. Slope case

	No.	Slope & height	Wind speed & roughness
Slope case	Case 1	*S*0.3*H*1.5	U_0=10m/s α =0.16
	Case 2	*S*0.4*H*1.5	
	Case 3	*S*0.5*H*1.5	
	Case 4	*S*0.6*H*1.5	
	Case 5	*S*0.75*H*1.5	
	Case 6	*S*1*H*1.5	

Table 2. Wind speed case

	No.	Slope & height	Wind speed (m/s)	Roughness
Wind speed case	Case 7	*S*1*H*1.5	10	α =0.16
	Case 8	*S*1*H*1.5	15	
	Case 9	*S*1*H*1.5	20	
	Case 10	*S*0.3*H*1.5	10	
	Case 11	*S*0.3*H*1.5	15	
	Case 12	*S*0.3*H*1.5	20	

Table 3. Height case

	No.	Slope & height	Wind speed & roughness
Height case	Case 13	$S0.5H1$	U_0=10m/s α =0.16
	Case 14	$S0.5H1.5$	
	Case 15	$S0.5H2$	

Table 4. Roughness case

	No.	Slope & height	Rough-ness	Wind speed
Roughness case	Case 16	$S1H1.5$	0.12	U_0=10m/s
	Case 17	$S1H1.5$	0.16	
	Case 18	$S1H1.5$	0.22	
	Case 19	$S1H1.5$	0.26	

(Slope: $S = H / L_1$, H1=100m, H1.5=150m, H2=200m, the roughness is presented as α .)

The flow field at the height 200m from the surface is selected for study, as shown in figure 3, ten measuring locations along the wind direction x are selected and defined as $x1$, $x2$...$x10$.

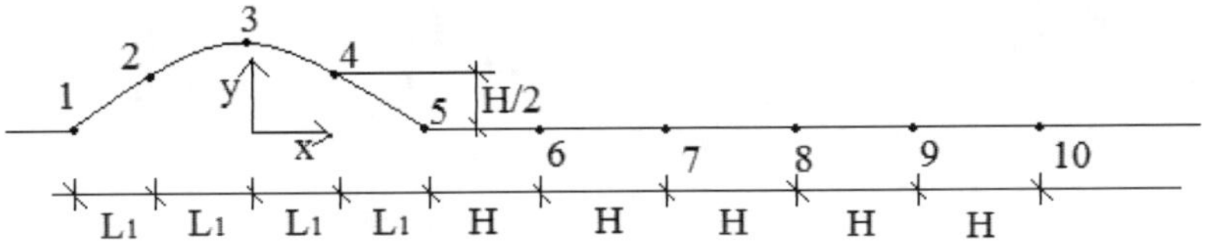

Figure 3. Measuring locations sketch

2.4 Wind speed-up ratio

The average wind speed at the height above the horizontal ground is defined as $U_0(z)$, and the speed at the corresponding height z above the mountain surface is defined as $U(z)$, and the speed acceleration ratio ΔS at the two heights is called acceleration effect and the formula is as follows:

$$\Delta S = \frac{U(z) - U_0(z)}{U_0(z)} \tag{2}$$

3. Results

3.1 Slope

In order to describe the changes and laws of wind speed under different slope conditions more accurately, the acceleration ratio was used, as shown in figure 4. In the figure, the horizontal axis represents the acceleration ratio and the vertical axis represents the vertical height from the surface. It can be seen from the figure that at location $x1$&$x2$, the influence of slope on wind speed is not particularly obvious. When airflow passes the top of the hill(location $x3$), it presents obvious acceleration effect. The larger the slope is, the more obvious the acceleration effect of wind speed near the surface will be. The maximum acceleration ratio reaches 0.7.

When the wind passes to the lee side (location $x5$→8), if the slope is greater than 0.5, the wind speeds along height at each locations shows the tendency of decrease then increase first, such as at the location $x5$, when the slope S=1, the wind speed starts from zero at the ground and then increased, following by reduce to the minimum when the height h is 50 m. It indicates that there is obvious eddy phenomenon near the foot of the leeward. For hills with slope S=0.5, the vortex disappears at 2H from the foot of the mountain (location $x7$), while for mountains with slope 1, the vortex disappears at 4H

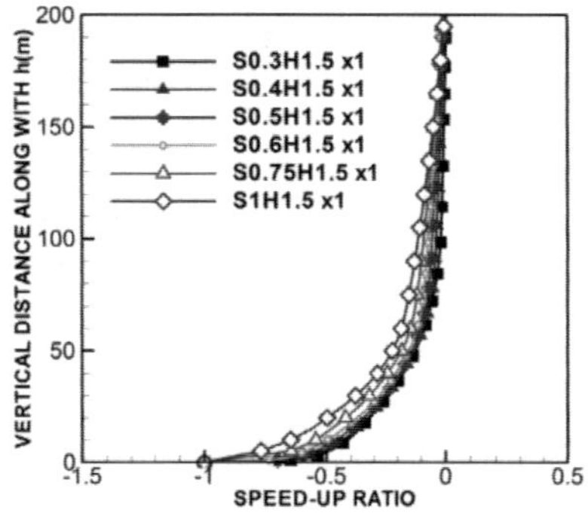

Location $x1$

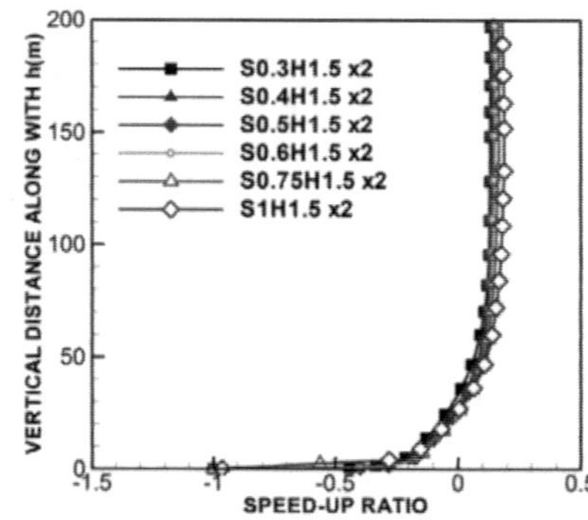

Location $x2$

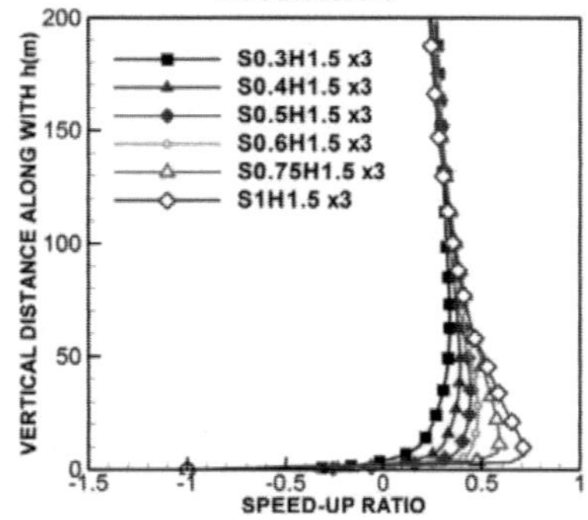

Location $x3$

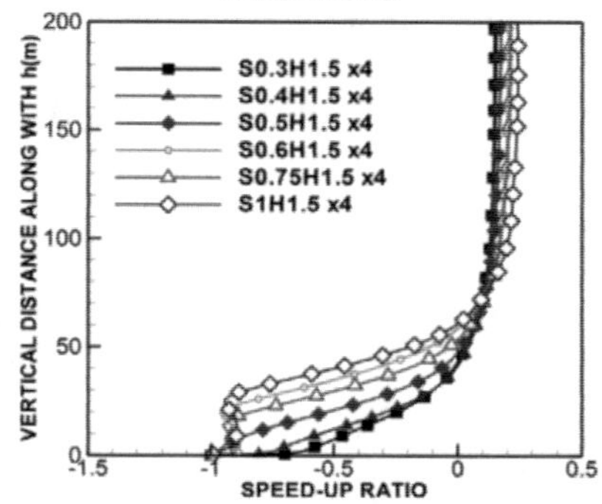

Location $x4$

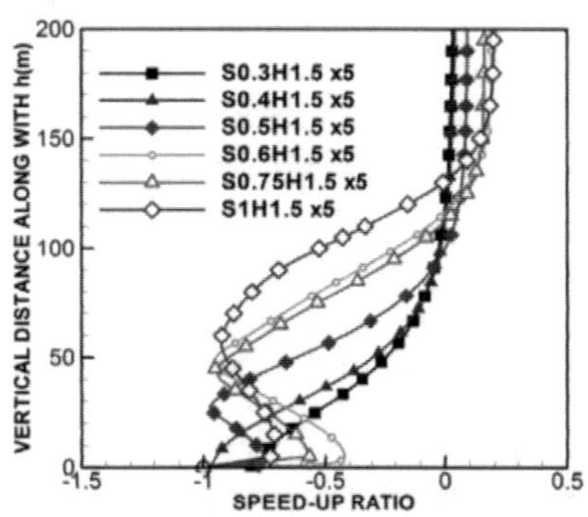

Location $x5$

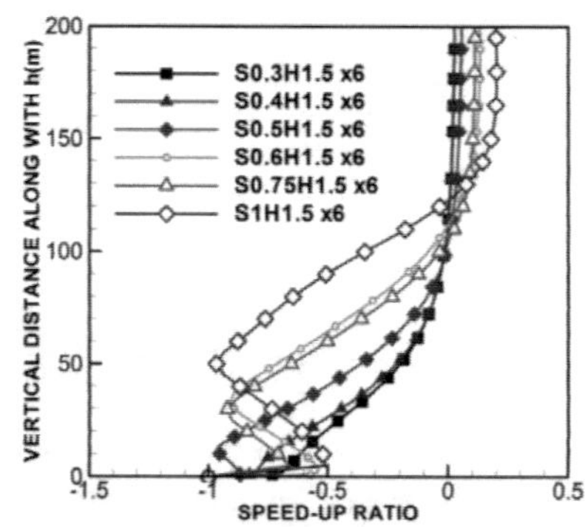

Location $x6$

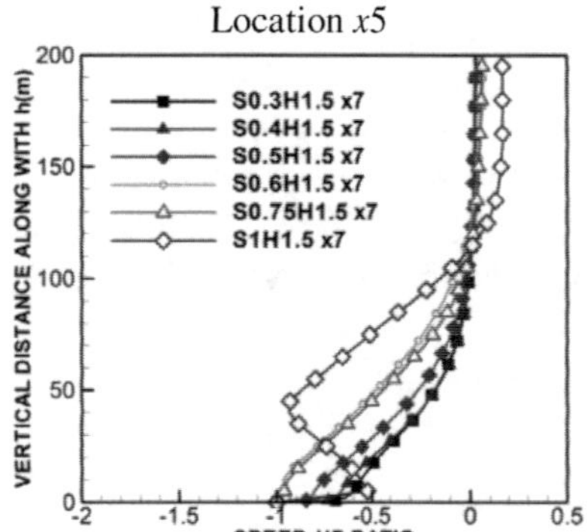

Location $x7$

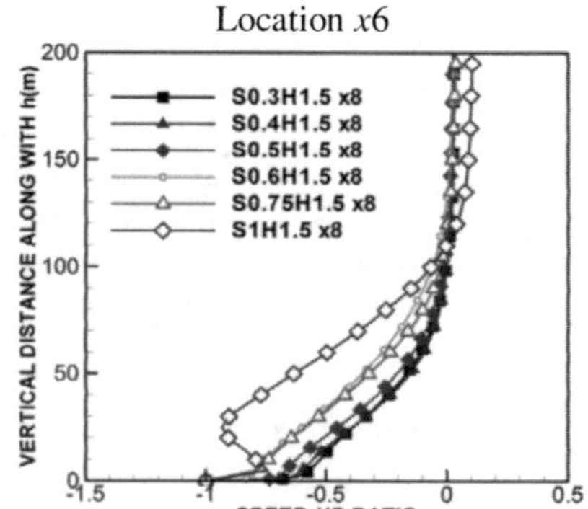

Location $x8$

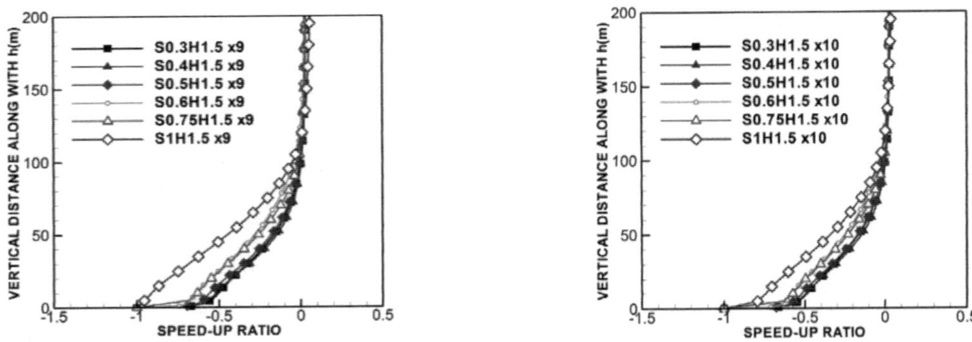

Location $x9$ Location $x10$

Figure 4. The speed-up ratios around the hill with different slopes

from the foot of the mountain (location x9).This indicates that the steeper the slope of the mountain, the farther the eddy range.

3.2 Wind speed

The analysis of the acceleration ratio change of the wind field on steep slope and gentle slope varying the incoming wind speed are depicted in figure 5.

For both the steeper hill(S=1) and gentle hill(S=0.3), the change of the incoming wind speeds has little influence on the wind speed distribution rules at location $x2$&3; but at the leeward sides(location $x5$), the wind profile with larger incoming speed referring to the gentle hill(S=0.3) presents slower recovery.

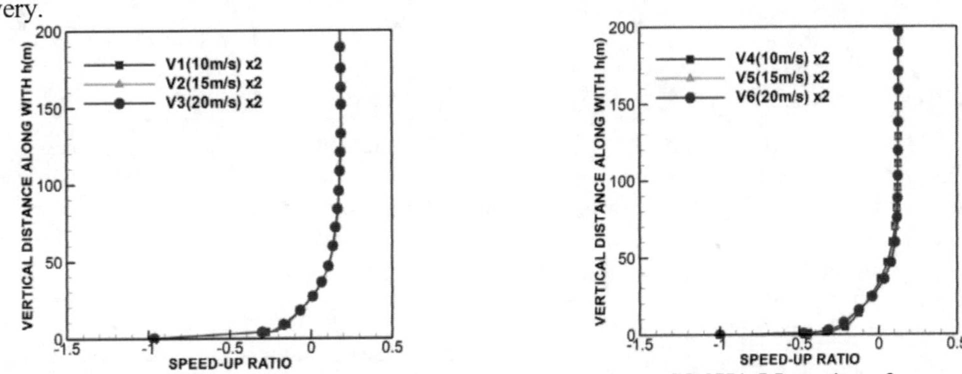

S1H1.5 Location $x2$ S0.3H1.5 Location $x2$

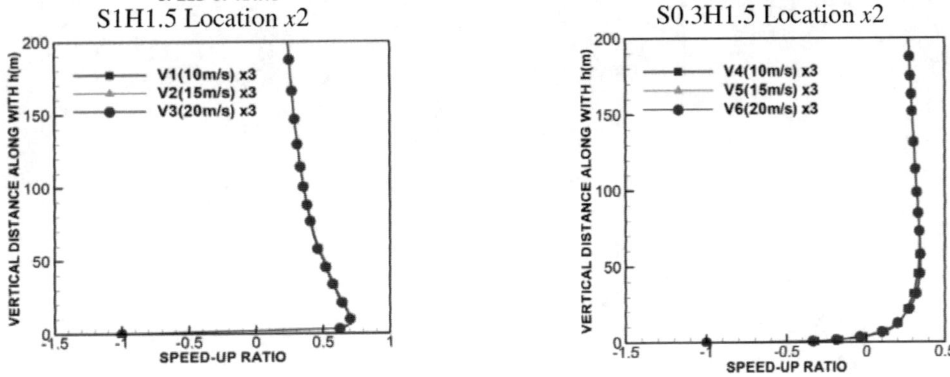

S1H1.5 Location $x3$ S0.3H1.5 Location $x3$

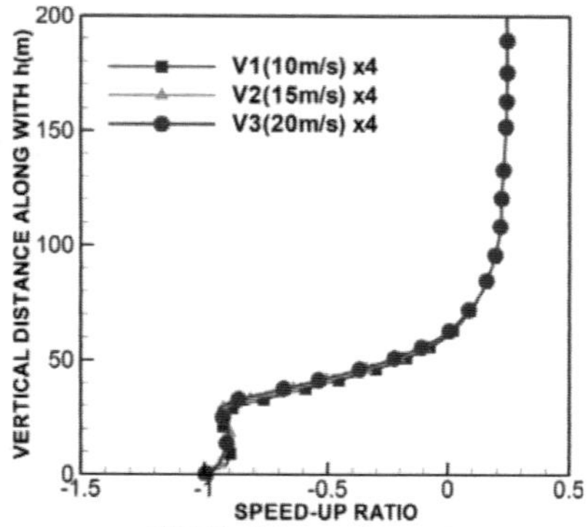

S1H1.5 Location $x4$

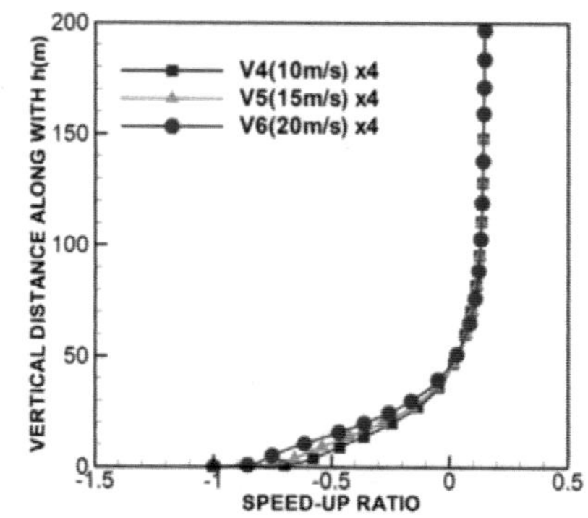

S0.3H1.5 Location $x4$

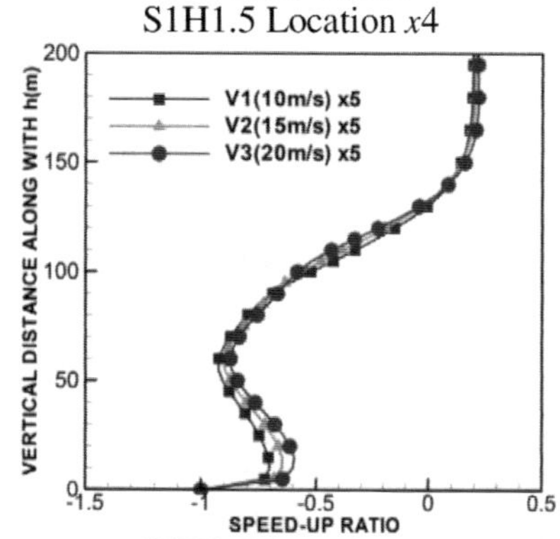

S1H1.5 Location $x5$

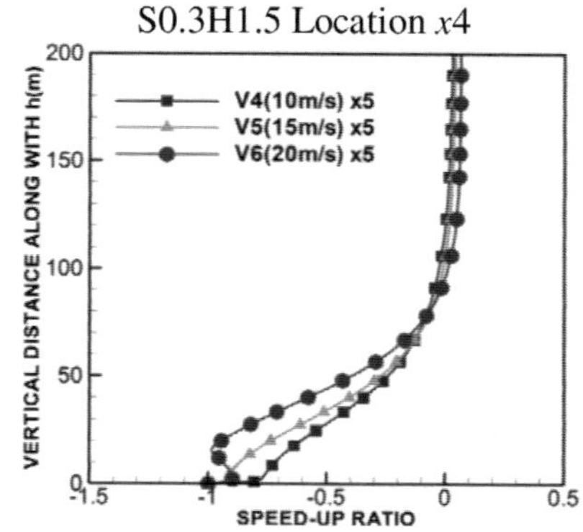

S0.3H1.5 Location $x5$

Figure 5. The speed-up ratios around the hill with different wind speeds(left: S1H1.5, right: S0.3H1.5)

3.3 Height

The hill models with different heights(H1=100m, H2=150m, H3=200m) are adopted for research, as shown in figure 6. The change of mountain height has little influence on the wind profile on the windward surface(location $x2$&3). At the leeward side of the mountain, when the hill height increase, the wind lessens notably near the ground, i.e. for the case S0.5H2 at location $x5$, the value of speed-up ratios are close to minus one within the range of 80m above the ground, which means that the wind velocity is very small.

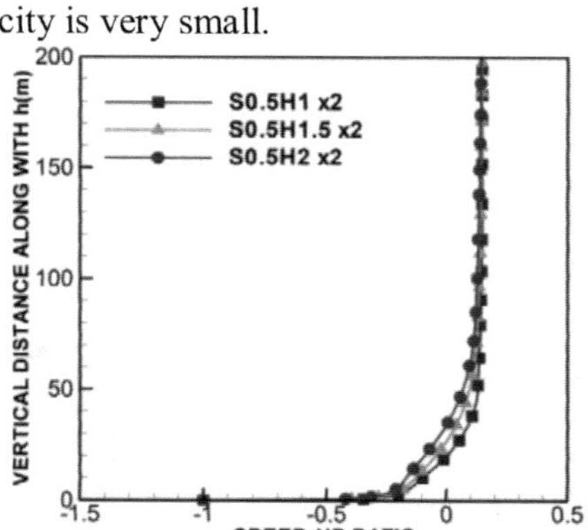

Location $x2$

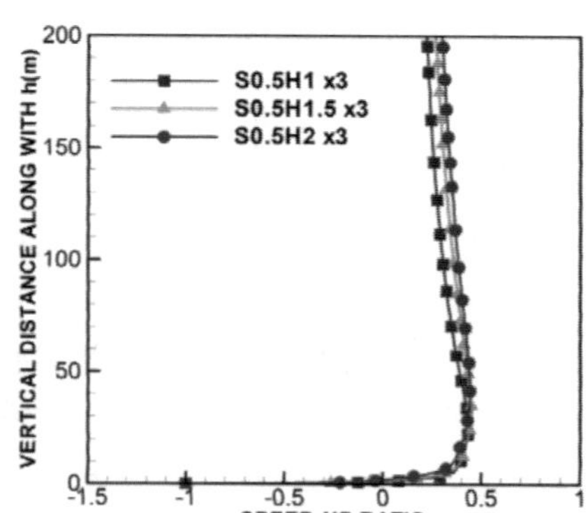

Location $x3$

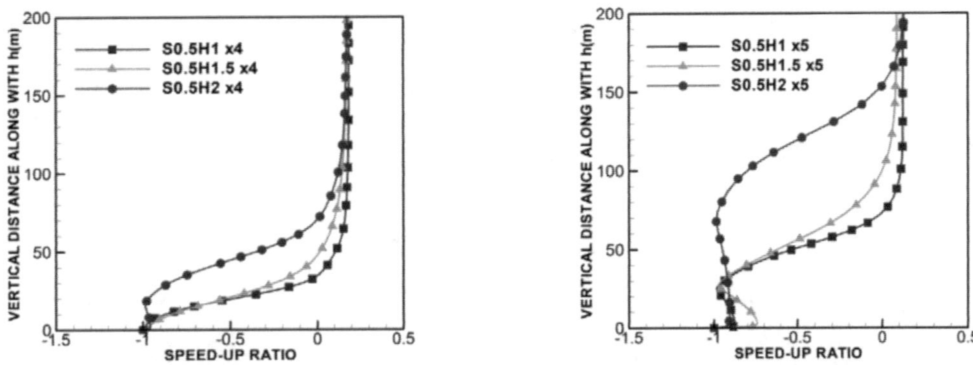

Location $x4$ Location $x5$

Figure 6. The speed-up ratios around the hill with different heights

3.4 Roughness

Figure 7 shows the distribution laws of wind speed profile around the hill affected by roughness. The change of roughness has little influence on the wind profile on the windward side (location x2&3). But for the leeward side, the effect of roughness is reflected. The speed-up ratio at the same height with larger roughness is smaller, that is, with the increase of roughness, the acceleration ratio will decrease correspondingly.

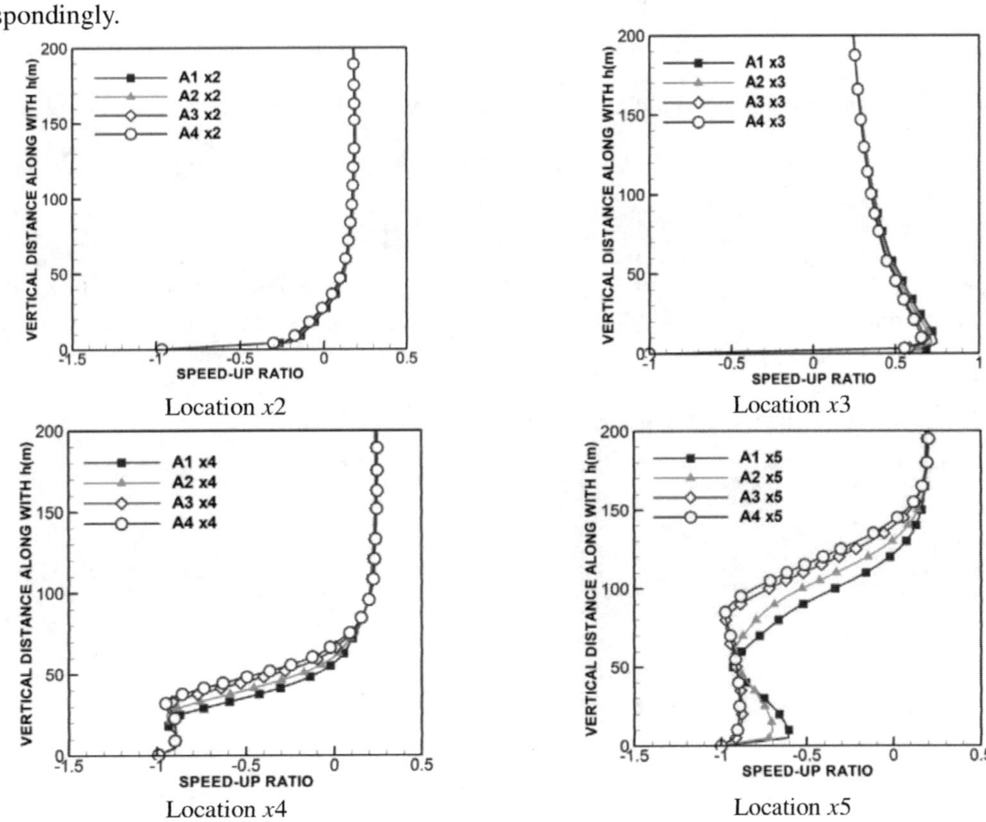

Location $x4$ Location $x5$

Figure 7. Influence of roughness(A1: α=0.12, A2: α=0.16, A3: α=0.22, A4: α=0.26)

4. Conclusions

Based on the numerical simulation of the hilly terrain, the conclusions can be drawn as follows:

The windward side of the hill: the influence of external factors on the wind speed rules from the hill foot to the hillside is small and negligible. For the range from the hillside to the top, the slope has a significant effect on the wind speed. Within the 100m height above the ground, when the hill is steeper, the wind speed is significantly enhanced.

The top of the hill: within 100m height above the ground, the wind speed's maximum increment is 1.5 times of the incoming value.

The leeward side of the hill: the factors, including the slope, height and roughness, have great influence on the wind environment. When the slope is greater than or equal to 0.5, the phenomenon of eddies present from the hillside to the foot on the leeward side. The increase of the height of the mountain weakened the wind speed on the leeward side. The wind speed weakness trend is also obvious for larger roughness.

5. References

[1] Kang Z., China debris flow research, Beijing: science press, 1989.
[2] Jackson P S , Hunt J C R . Turbulent wind flow over a low hill[J]. Quarterly Journal of the Royal Meteorological Society, 2010, 101(430):929-955.
[3] Gong W , Ibbetson A . A wind tunnel study of turbulent flow over model hills[J]. Boundary-Layer Meteorology, 1989, 49(1-2):113-148.
[4] Miller C A, Davenport A G. Guidelines for the calculation of wind speed-ups in complex terrain[J]. Journal of Wind Engineering & Industrial Aerodynamics, 1998, s 74 – 76(2):189-197.
[5] Uchida T , Ohya Y . Large-eddy simulation of turbulent airflow over complex terrain[J]. Journal of Wind Engineering and Industrial Aerodynamics, 2003, 91(1-2):219-229.
[6] Kondo K, Tsuchiya M, Sanada S. Evaluation of effect of micro-topography on design wind velocity[J]. Journal of Wind Engineering & Industrial Aerodynamics, 2002, 90(12):1707-1718.
[7] Yang Z, Xue H, Yuan C. Evaluation of wind energy resources in complex terrain [C], Annual conference of Chinese solar energy society. 2003
[8] Load Code for the Design of Building Structures. GB50009-2001, Beijing: China Architecture & Building Press, 2002.
[9] Zhao Y, CFD simulation of wind field in complex terrain [D], North China Electric Power University (Beijing), 2011.

Acknowledgement

The authors would like to acknowledge with great gratitude for the supports of the national science foundation of China(grant No: 51778551), the science and technology plan project of Xiamen(grant No: 3502Z20161016), the young teachers education science research project of Fujian(grant No: JAT170410), the Natural Science Foundation of Fujian Province(grant No. 2018J01525), and the projects of Xiamen University of Technology(grant No. XPDKQ18034 & XPDKQ18042).

Stability Analysis of a Model for Phenol and Cresol Mixture Degradation

N Dimitrova[1] and P Zlateva[1, 2]

1 Institute of Mathematics and Informatics, BAS, Sofia, Bulgaria
2 Institute of Robotics, Bulgarian Academy of Sciences, Sofia, Bulgaria
E-mail: nelid@bio.bas.bg, plamzlateva@abv.bg

Abstract. We propose a mathematical model for phenol and p-cresol mixture degradation in a continuously stirred bioreactor. The model is described by three nonlinear ordinary differential equations. The novel idea in the model design is the biomass specific growth rate, known as sum kinetics with interaction parameters (SKIP). We establish existence and uniform boundedness of positive solutions. Then we determine the equilibrium points of the model and study their asymptotic stability. The theoretical investigations are based on the classical dynamic systems theory. Numerical simulations are included to confirm and fill out the theoretical studies.

1. Introduction

It is well known that aromatic compounds are heavy environmental pollutants. Wastewaters from petroleum refineries, coal mining and a variety of industrial chemical syntheses contain many aromatics such as phenol, cresols, nitrophenols, etc. The biodegradation of one or all components depends on the composition of the particular mixture and the used microorganisms [1], [2], [3]. The adequate analysis of interactions between the compounds and their influence on microbial growth is very important for understanding the simultaneous metabolism of phenolic mixtures.

The availability of appropriate mathematical models for these biochemical processes is essential for quantification and estimation of interactions between the substrate components influencing the growth and degradation behavior of microorganisms. A few mathematical models have been developed to describe the mixed substrate biodegradation [4], [5], [6].

Recent practical experiments support the design of the so-called SKIP models (the *sum kinetics with interaction parameters*) for substrate mixture degradation, because they describe significantly better the microbial growth [7], [8].

Here, we perform a stability analysis of a SKIP model in a continuously stirred bioreactor, describing the mutual influence of phenol and 4-methylphenol (p-cresol).

The paper is organized as follows. The next Section 2 includes a short description of the proposed mathematical model. Section 3 is devoted to investigating the properties of the model solutions, including steady states analysis. The last Section 4 proposes numerical simulations and visualization of the model dynamics.

2. Model description

We propose the following model for phenol and p-cresol mixture degradation in a continuously stirred bioreactor

Content from this work may be used under the terms of the Creative Commons Attribution 3.0 licence. Any further distribution of this work must maintain attribution to the author(s) and the title of the work, journal citation and DOI.
Published under licence by IOP Publishing Ltd

$$\frac{dX(t)}{dt} = \left(\mu(S_{ph},S_{cr}) - D\right)X(t) \tag{1}$$

$$\frac{dS_{ph}(t)}{dt} = -k_{ph}\mu(S_{ph},S_{cr})X(t) + D\left(S_{ph}^{0} - S_{ph}(t)\right) \tag{2}$$

$$\frac{dS_{cr}(t)}{dt} = -k_{cr}\mu(S_{ph},S_{cr})X(t) + D\left(S_{cr}^{0} - S_{cr}(t)\right), \tag{3}$$

where $\mu(S_{ph},S_{cr})$ is the specific growth rate, presented by

$$\mu(S_{ph},S_{cr}) = \frac{\mu_{\max(ph)}S_{ph}}{k_{s(ph)} + S_{ph} + I_{cr/ph}S_{cr}} + \frac{\mu_{\max(cr)}S_{cr}}{k_{s(cr)} + S_{cr} + I_{ph/cr}S_{ph}} \tag{4}$$

The model function $\mu(S_{ph},S_{cr})$ accounts for the mixture effects of phenol and p-cresol compounds and is known as *sum kinetics with interaction parameters* (SKIP) [7], [9].

The definition of the state variables X, S_{ph} and S_{cr} as well as of the model parameters is given in Table 1. All coefficients are assumed to be positive.

As usual, the dilution rate D is considered as a control variable.

Table 1. Model variables and parameters.

	Definitions	Values
X	biomass concentration [g/dm^3]	-
S_{ph}	phenol concentration [g/dm^3]	-
S_{cr}	p-cresol concentration [g/dm^3]	-
D	dilution rate [h^{-1}]	-
S^0_{ph}	influent phenol concentration [g/dm^3]	0.7
S^0_{cr}	influent p-cresol concentration [g/dm^3]	0.3
k_{ph}	metabolic coefficient [S_{ph}/X]	11.7
k_{cr}	metabolic coefficient [S_{cr}/X]	5.8
$I_{ph/cr}$	interaction coefficient indicating the degree to which phenol affects the p-cresol biodegradation	0.3
$I_{cr/ph}$	interaction coefficient indicating the degree to which p-cresol affects the phenol biodegradation	8.6
$\mu_{max(ph)}$	maximum specific growth rate on phenol as a single substrate [h^{-1}]	0.23
$\mu_{max(cr)}$	maximum specific growth rate on p-cresol as a single substrate [h^{-1}]	0.17
$k_{s(ph)}$	saturation constant for cell growth on phenol [g/dm^3]	0.11
$k_{s(cr)}$	saturation constant for cell growth on p-cresol [g/dm^3]	0.35

3. Stability analysis of the model

3.1. Basic properties of the model solutions

Define the set

$$\Omega = \left\{(X,S_{ph},S_{cr}):\ X > 0, S_{ph} > 0, S_{cr} > 0\right\}.$$

Theorem 1. For any starting point $\left(X(0), S_{ph}(0), S_{cr}(0)\right) \in \Omega$ the solution $\left(X(t), S_{ph}(t), S_{cr}(t)\right)$ of system (1)-(3) is well defined and uniformly bounded for each $t \geq 0$.

Proof. It follows from (1) that if point $X(\tau) = 0$ for some $\tau \geq 0$ then $\frac{dX}{dt}(\tau) = 0$. The uniqueness of the solution implies $X(t) = 0$ for all $t \geq \tau$. Therefore it is enough to consider only positive initial conditions implies $X(t) > 0$.

Using equation (1) we obtain consecutively

$$\frac{dX}{dt} = \int_0^t \left(\mu(S_{ph}(\tau), S_{cr}(\tau)) - D\right) d\tau$$

$$X(t) = X(0)e^{-\int_0^t (\mu(S_{ph}(\tau), S_{cr}(\tau)) - D) d\tau} > 0 \quad \text{for each } t \geq 0.$$

Equations (1) and (2) imply

$$\frac{dS_{ph}}{dt} + k_{ph}\frac{dX}{dt} = D\left(S_{ph}^0 - (S_{ph} + k_{ph}X)\right)$$

Denoting $Z_1(t) = S_{ph}(t) + k_{ph}X(t)$ we obtain

$$\frac{dZ_1}{dt}(t) = D\left(S_{ph}^0 - Z_1(t)\right)$$

$$Z_1(t) = S_{ph}^0 + \left(Z_1(0) - S_{ph}^0\right)e^{-Dt} > 0 \quad \text{for all } t \geq 0.$$

Therefore,

$$\max\{Z_1(0), S_{ph}^0\} \geq Z_1(t) \geq \min\{Z_1(0), S_{ph}^0\} > 0 \quad \text{for all } t \geq 0.$$

Similarly, using equations (1) and (3) and denoting $Z_2(t) = S_{cr}(t) + k_{cr}X(t)$ we get

$$\frac{dZ_2}{dt}(t) = D(S_{cr}^0 - Z_2(t)),$$

$$Z_2(t) = S_{cr}^0 + (Z_2(0) - S_{cr}^0)e^{-Dt} > 0 \quad \text{for all } t \geq 0,$$

$$\max\{Z_2(0), S_{cr}^0\} \geq Z_2(t) \geq \min\{Z_2(0), S_{cr}^0\} > 0 \quad \text{for all } t \geq 0.$$

The above presentations mean that the model solutions $\left(X(t), S_{ph}(t), S_{cr}(t)\right)$ are uniformly bounded and thus exist for all time $t \geq 0$. This completes the proof of Theorem 1.

Remark 1. In the proof of Theorem 1 we didn't use the explicit expression of $\mu\left(S_{ph}, S_{cr}\right)$ from (4), thus the assertion is valid for any continuously differentiable and bounded specific growth rate function function $\mu(.)$.

3.2. Equilibrium points of the model

The equilibrium points of (1)-(3) are solutions of the following system of algebraic equations

$$\left(\mu\left(S_{ph}, S_{cr}\right) - D\right)X = 0 \tag{5}$$

$$-k_{ph}\mu\left(S_{ph}, S_{cr}\right)X + D\left(S_{ph}^0 - S_{ph}\right) = 0 \tag{6}$$

$$-k_{cr}\mu\left(S_{ph}, S_{cr}\right)X + D(S_{cr}^0 - S_{cr}) = 0. \tag{7}$$

IOP Conf. Series: Earth and Environmental Science **356** (2019) 012009 doi:10.1088/1755-1315/356/1/012009

Obviously, the point $E_0 = (0, S_{ph}, S_{cr})$ (within $X = 0$) is always an equilibrium point of the model. We are looking now for solutions of (5)-(7) assuming that $X \neq 0$).

From (7) we obtain

$$X = \frac{D(S_{cr}^0 - S_{cr})}{k_{cr}\mu(S_{ph}, S_{cr})}.$$

Obviously, $X > 0$ iff $S_{cr} < S_{cr}^0$ holds. Substituting the above expression of X into equation (6) implies

$$S_{ph} = S_{ph}^0 - \frac{k_{ph}}{k_{cr}}(S_{cr}^0 - S_{cr}).$$

Denote for simplicity

$$K = \frac{k_{ph}}{k_{cr}}, \quad S^0 = S_{ph}^0 - KS_{cr}^0$$

and make the following assumption

$$S^0 > 0. \tag{8}$$

Then the presentation $S_{ph} = S^0 + KS_{cr}$ holds true and substituting the latter expression in (5) we obtain an equation with respect to S_{ph} only:

$$\frac{\mu_{\max(ph)}(S^0 + KS_{cr})}{k_{s(ph)} + S^0 + KS_{cr} + I_{cr/ph}S_{cr}} + \frac{\mu_{\max(cr)}S_{cr}}{k_{s(cr)} + S_{cr} + I_{ph/cr}(S^0 + KS_{cr})} = D.$$

Let us introduce the notations

$$A_1 = 1 + KI_{ph/cr}, \quad A_2 = K + I_{cr/ph}, \quad A_3 = k_{s(cr)} + I_{ph/cr}S^0, \quad A_4 = k_{s(ph)} + S^0$$

and further

$$A = \mu_{\max(ph)}KA_1 + \mu_{\max(cr)}A_2 - DA_1A_2$$

$$B = \mu_{\max(ph)}(S^0A_1 + KA_3)\mu_{\max(cr)}A_4 - D(A_1A_4 + A_2A_3)$$

$$C = (\mu_{\max(ph)}S^0 - DA_4)A_3.$$

Then the S_{cr}-component of the equilibrium point should satisfy the quadratic equation

$$AS_{cr}^2 + BS_{cr} + C = 0, \tag{9}$$

which coefficients A, B, C depend on the model parameter D. To check the existence of real (and positive) roots of (9), we need to calculate its discriminant $\Delta(D) = B^2 - 4AC$. Straightforward calculations lead to the following expression for $\Delta(D)$.

$$\Delta(D) = aD^2 + bD + c,$$

where

$$a = (A_1A_4 - A_2A_3)^2 > 0,$$

$$b = 2\mu_{\max(ph)}(A_1A_2A_3S^0 - A_1^2A_4S^0 - A_3(A_1A_4 + A_2A_3)K + 2A_1A_3A_4K) +$$
$$+2\mu_{\max(cr)}A_4(A_2A_3 - A_1A_4),$$

$$c = \left(\mu_{\max(ph)} A_1 S^0 + \mu_{\max(cr)} A_4\right)^2 + \mu^2_{\max(ph)} K A_3 (K A_3 - 2 A_1 S^0) +$$
$$+ 2\mu_{\max(ph)} \mu_{\max(cr)} A_3 (K A_4 - 2 A_2 S^0).$$

Obviously, the model can possess up to two internal (with positive components) equilibrium points. Their existence depends on the coefficient values as well as on values of D. More precisely, we have to solve the inequality $aD^2 + bD + c \geq 0$ with respect to D to determine bounds of D, for which one or two positive equilibria do exist. This will be done numerically in the next section, using the values of the model coefficients, given in the last column in Table 1. It is worth to note that these values are determined by real-life laboratory experiments [9].
Before that we shall study some important properties of the model solutions.
In what follows we shall use the notation

$$D^+ = \mu\left(S^0_{ph}, S^0_{cr}\right).$$

Theorem 2. The dynamic system (1)-(3) undergoes a transcritical bifurcation at the equilibrium point $E_0 = \left(0, S^0_{ph}, S^0_{cr}\right)$ for $D = D^+$.
Proof. Let $J = (E_0; D^+)$ be the Jacobian of the model equations (1)-(3) evaluated at $E_0 = \left(0, S^0_{ph}, S^0_{cr}\right)$ and $D = D^+$. It is straightforward to see that the eigenvalues of $J = (E_0; D^+)$ are the roots of the following characteristic equation $\lambda\left(\lambda^2 + k_{cr} D^+ \lambda - k_{ph} D^{+^2}\right) = 0$; the latter obviously possesses one root $\lambda = 0$ and two more real roots with opposite signs. This means that E_0 is a nonhyperbolic equilibrium, and a transcritical bifurcation occurs at $D = D^+$ [10].
Remark 2. It is known [10] that the occurrence of a bifurcation for some parameter values (in our case for $D = D^+$ leads to destabilization of the model dynamics. It is not difficult to see that if $D > D^+$ then E_0 remains the unique equilibrium of (1)-(3) and it is necessarily the global attractor. This means that for any initial conditions $\left(X(0), S_{ph}(0), S_{cr}(0)\right) \in \Omega$, the relations $X(t) \to 0$ and $S_{ph}(t) \to S^0_{ph}$, $S_{cr}(t) \to S^0_{cr}$ hold as $t \to +\infty$. The practical interpretation of this fact is that the organisms are washed-out and the environment is not detoxified, which is the worst possible case.
For the practical applications, the most important equilibrium point is that one with positive components, i. e. the internal equilibrium. The existence of such an equilibrium, and especially when it is asymptotically stable, means that the organisms remain in the system and detoxification is carried out.
Theorem 3. If $D < D^+$ and there exists at least one internal equilibrium $E^* = \left(X^*, S^*_{ph}, S^*_{cr}\right) > 0$, then there exists time $T > 0$, such that $S_{ph}(t) < S^0_{ph}$ and $S_{cr}(t) < S^0_{cr}$ hold true for all $t > T$.
Proof. The existence of the equilibrium point E^* means that there exists some $D^+ \in (0, D^+)$ such that $\mu\left(S^*_{ph}, S^*_{cr}\right) = D^*$ is satisfied. Assume that $S_{ph}(t) \geq S^0_{ph}$ and $S_{cr}(t) \geq S^0_{cr}$ hold for each $t > 0$. Then we have from (2) that

$$\frac{dS_{ph}(t)}{dt} = -k_{ph}\mu\left(S_{ph}, S_{cr}\right)X(t) + D\left(S^0_{ph} - S_{ph}(t)\right) < 0$$

Barbălat's Lemma [11] implies that

$$0 = \lim_{t\to\infty} \frac{dS_{ph}(t)}{dt} = \lim_{t\to\infty}\left(-k_{ph}\mu\left(S_{ph}, S_{cr}\right)X(t) + D\left(S^0_{ph} - S_{ph}(t)\right)\right)$$

which leads to $S_{ph}(t) \to S^0_{ph}$ and $X(t) \to 0$ as $t \to \infty$. Similarly, using equation (3), we obtain $S_{cr}(t) \to S^0_{cr}$ and $X(t) \to 0$ as $t \to \infty$. Further we have that $\mu\left(S^*_{ph}, S^*_{cr}\right) = D^* < D^+ = \mu\left(S^0_{ph}, S^0_{cr}\right)$. The continuity of $\mu(\cdot)$ and the relations $S_{ph}(t) \to S^0_{ph}$ and $S_{cr}(t) \to S^0_{cr}$ as $t \to \infty$ imply that there exists $\nu > 0$ such that

$$\mu(S_{ph}(t),S_{cr}(t)) - D^* = \mu(S_{ph}(t),S_{cr}(t)) - \mu(S_{ph}^*,S_{cr}^*) \geq \nu$$

for all sufficiently large t. It follows then that $\dot{X}(t) = (\mu(S_{ph}(t),S_{cr}(t)) - D^*)X(t) \geq \nu X(t)$ for all sufficiently large t, which contradicts the boundedness of $X(t)$. Hence, there exists a sufficiently large $T_1 > 0$ with $S_{ph}(T_1) \leq S_{ph}^0$. Moreover, if the equality $S_{ph}(T_1) = S_{ph}^0$ holds true, then we have

$$\frac{dS_{ph}(t)}{dt} = -k_{ph}\mu\left(S_{ph}(T_1),S_{cr}(T_1)\right)X(T_1) + D\left(S_{ph}^0 - S_{ph}(T_1)\right) =$$

$$= -k_{ph}\mu\left(S_{ph}(T_1),S_{cr}(T_1)\right)X(T_1) < 0\,.$$

The last inequality shows that $S_{ph}(t) < S_{ph}^0$ for each $t > T_1$. Similarly, one can show that there exists $T_2 > 0$ such that $S_{cr}(t) < S_{cr}^0$ for each $t > T_2$. Taking $T \geq \max\{T_1, T_2\}$ proves the theorem.

4. Numerical simulation

Using the numerical values in the last column of Table 1, we obtain the following results, which confirm and fill out the theoretical studies from the previous section. We have

$$D^+ = \mu(S_{ph}^0, S_{cr}^0) = 0.1068$$

and further

$$\Delta(D) \geq 0 \quad \text{if} \quad D \geq 0.07947 =: D^-\,.$$

Therefore,

(*i*) if $D \in (D^-, D^+)$ then there exist two positive equilibrium points, E_1 and E_2, of the model. $D = D^+$ is a bifurcation value, leading to $X = 0$.

(*ii*) if $D = D^-$ then $\Delta(D^-) = 0$ and the two positive equilibria coalesce, i. e. $E_1 \equiv E_2$. Hence, for $D \in (0, D^-)$ the point E_0 is the unique equilibrium of the model.

(*iii*) if $D \geq D^+$ then E_0 is the unique equilibrium point of the model as a result of bifurcation for $D = D^+$.

It follows from (*i*) - (*iii*) that the practically useful values of the control variable D belong to the interval (D^-, D^+). The boundedness of $D \in (D^-, D^+)$ is not restrictive in practical applications, since the dilution rate D is proportional to the speed of the pumping mechanism which feeds the bioreactor. Thus there always exist a lower and an upper bound for D [12].
We shall demonstrate below the dynamic behavior of the model solutions on a particular numerical example.

Example. $D = 0.1 \in (D^-, D^+)$
In this case there exist two positive (internal) equilibrium points

$$E_1 = (0.05126, 0.1002, 0.002687) \quad \text{and} \quad E_2 = (0.01359, 0.5411, 0.2212),$$

such that E_1 is locally unstable (a saddle), E_2 is a locally asymptotically stable equilibrium point (a node), and E_0 is locally unstable steady state (a saddle).

The left plot in Figure 1 shows the solutions $X(t)$, $S_{ph}(t)$, $S_{cr}(t)$; there, the dash-lines pass through the corresponding components of the stable equilibrium E_2. The right plot in Figure 1 as well as the two plots in Figure 2 illustrate the trajectories of the model in different phase planes with two different initial points. In these figures, the solid circles denote the corresponding components of the stable equilibrium E_2, the boxes denote the corresponding components of the unstable steady states E_0 and E_1.

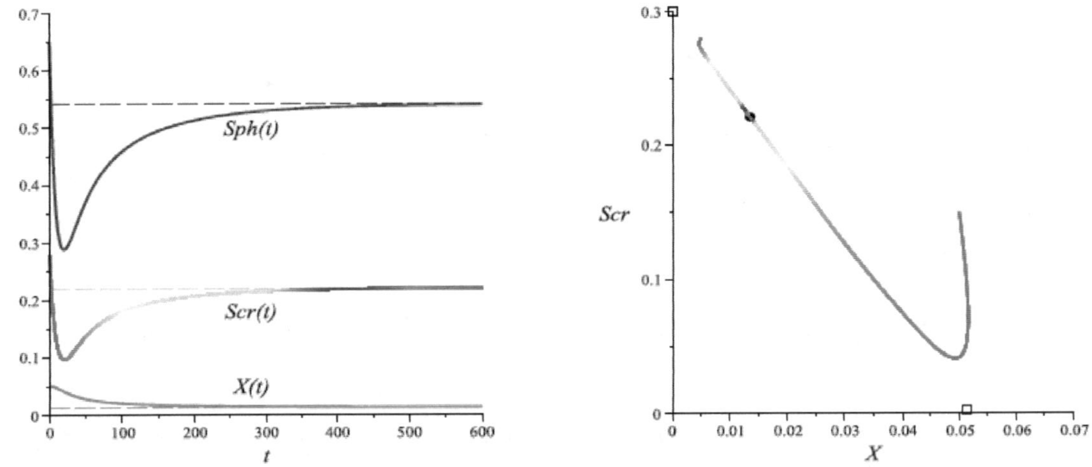

Figure 1. Time evolution of the model solutions (left); trajectories in the (*X; Scr*)-plane (right).

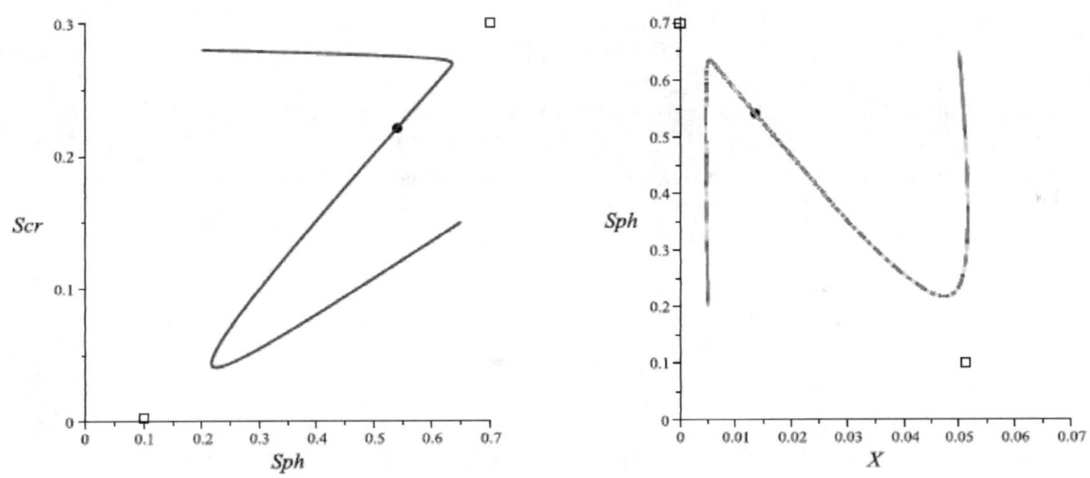

Figure 2. Trajectories in the (*Sph; Scr*)-plane and in the (*X; Sph*)-plane.

5.References

[1] Sharma N K, Philip L and Murty B S 20102 Aerobic degradation of phenolics and aromatic hydrocarbons in presence of cyanide, *Biresour. Technol.* **121** pp 263-273

[2] Tomei M C and Annesini M C 2008 Biodegradation of phenolic mixtures in a sequencing batch reactor, *A kinetic study, Env. Sci. Pollut. Res.* **15** (3) pp 188-195

[3] Yemendzhiev H, Zlateva P and Alexieva Z 2012 Comparison of the biodegradation capacity of two fungal strains toward a mixture of phenol and cresol by mathematical modeling, *Biotechnology and Biotechnological Equipment* **26** (5) pp 3278-3281

[4] Angelucci D M, Annesini M C and Tomei M C 2013 Modelling of biodegradation kinetics for binary mixtures of substituted phenols in sequential bioreactors, *Chemical Engineering Transaction—AIDIC* **32** pp 1081-1086

[5] Datta A, Philip L and Bhallamudi S M 2014 Modeling the biodegradation kinetics of aromatic and aliphatic volatile pollutant mixture in liquid phase, *Chemical Engineering Journal* **241** pp 288-300

[6] Lepik R and Tenno T 2011 Biodegradability of phenol, resorcinol and 5-methylresorcinol as single and mixed substrates by active sludge, *Oil Shale* **28** (3) pp 425-446

[7] Reardon K F, Mosteller D C, Rogers J B, DuTeau N M and Kim K-H 2002 Biodegradation Kinetics of Aromatic Hydrocarbon Mixtures by Pure and Mixed Bacterial Cultures, *Environmental Health Perspectives* **110** (6) pp 1005-1011

[8] Hazrati H, Shayegan J and Seyedi S M 2015 Biodegradation kinetics and interactions of styrene and ethylbenzene as single and dual substrates for a mixed bacterial culture, *Journal of Environmental Health Science & Engineering* **10** 12 papes

[9] Yemendzhiev H, Gerginova M, Zlateva P, Stoilova I, Krastanov A and Alexieva Z 2008 Phenol and cresol mixture degradation by *Aspergillus awamori* strain: Biochemical and kinetic substrate interactions, *Proc. ECOpole* **2** (1) pp 153-159

[10] Wiggins S 1990 *Introduction to applied nonlinear dynamical systems and chaos*. (New York: Springer)

[11] Gopalsamy K 1992 *Stability and oscillations in delay differential equations of population dynamics*, (Dordrecht: Kluwer Academic Publishers)

[12] Grognard F and Bernard O 2006 Stability analysis of a wastewater treatment plant with saturated control, *Water Science Technology* **53** pp 149-157

Acknowledgments

The work of the first author has been partially supported by the National Scientific Program “Information and Communication Technologies for a Single Digital Market in Science, Education and Security (ICT in SES)", contract No D01-205/23.11.2018, financed by the Ministry of Education and Science in Bulgaria. The work of the second author has been accomplished with the financial support of the Ministry of Education and Science by Grant No DO1-161/28.08.2018 for the National Geo-Information Centre - part of the Bulgarian National Roadmap for Research Infrastructure.

Joint prediction of CO, NOx, NMHC pollutant concentrations in urban area

JL Zhao[1], MX Sun[2], Q Li[3]

1 Hong-Kong Baptist University, Kowloon Tong, Hong Kong
2 University of Essex, Colchester, UK
3 Duke University, NC, USA
E-mail: 1921938704@qq.com

Abstract. Primary air pollutants could directly and indirectly (through generating secondary air pollutants) threaten natural or human systems. In particular, urban air pollution issue becomes more and more significant in the recent decades, as a result of rapid urbanization. However, the estimation of multiple-pollutant concentrations is limited by high spatial and temporal variations, which hinders the accuracy of mechanistic modeling of air pollution. In this study, we employed an Artificial Neural Network (ANN) model to jointly predict multiple primary pollutants, including CO, NOx, and nan-methane hydrocarbons (NMHC), over the urban area of Italy. The results showed that performances of the ANN model (MSE and Pearson correlation) in joint prediction cross multiple pollutants were much better than in prediction of any single pollutant individually, indicating the joint measurements of multiple pollutants could favor the machine-learning model by providing useful information from one pollutant to predict another.

1.Introduction

Air pollution occurs when the air is contaminated by harmful chemical, physical or biological agents from natural or human sources. In the aspect of natural resource, dust from dry land, gas emissions from forest fires, ash from volcano eruptions and so on could significantly contribute to air pollution. Anthropogenic emissions through burning fossil fuel and biomass could also negatively affect the air quality. Air pollutants can be categorized into primary pollutants and secondary pollutants. Primary pollutants are the harmful substances directly entered into the air, including e.g., Carbon monoxide (CO), Nitrogen Oxides (NO_x), Nanmethane hydrocarbons (NMHC), and others. However, secondary pollutants are formed when primary pollutants react with atmospheric chemicals.

CO is colorless and odorless pollutant, which is produced by incomplete burning of fuels [1]. The primary source of CO is from motor vehicle exhausts. Thereby, CO usually accumulates in areas with heavy traffic. CO diffuses into human's blood and impair the ability of hemoglobin to carry oxygen in red blood cells, which can result vital organs failure, such as brain. NOx is a common mixture pollutants over the urban area pollutants, includes nitrogen monoxide (NO) and nitrogen dioxide(NO_2).Vehicle exhausts, industrial activities, and electricity producing contribute to the high concentration of NO_x. For example, NO_x are often generated when fuel is burned at a high temperature. NO_x could be extremely harmful when it dissolved into water in the atmosphere and from acid rain, which damages buildings, threatens the existence of plants and animals including human [2] [3]. Non-methane hydrocarbons (NMHCs) are long-lived chemical species in the atmosphere. They play an important role in regulating the air quality via reacting with hydroxyl radical (OH) and producing many harmful oxygenic compounds, and result in the formation secondary air

Content from this work may be used under the terms of the Creative Commons Attribution 3.0 licence. Any further distribution of this work must maintain attribution to the author(s) and the title of the work, journal citation and DOI.
Published under licence by IOP Publishing Ltd

pollutants in the atmosphere [4]. Vehicular emissions, biofuel combustion, biomass burning and industrial emissions were the major contributors to NMHCs.

Air those three major pollutants of interest could induce adversely short and long term effect on human healthy depending on the degree of concentration, length of exposure, type of pollutants and individual characteristics. In order to better understanding the complex process of dispersion of air pollutants and to make the connection between concentrations and the sources, models have been widely used to project the pollutant concentrations at present day and also in the future [5] including regional photochemical grid models and local-scale dispersion models[6]. For example, regional photochemical grid models are able to directly simulate the transport and formation of pollutants with atmospheric chemical reactions (i.e., Community Multiscale Air Quality(CMAQ) model). These models are computationally expensive and have been used at regional scale, but are not easily applicable to local-scale dynamics [6]. The date- driven models such Machine Learning (ML) models provide a promising way to learn purely from existing data and based on which generate computationally efficient prediction in the future. In this study, our objective is to accurately predict concentrations of multiple air pollutants including CO, NOx , and NMHCs.

2.Methodology

We used observations of multiple pollutants from Italy, with hourly temporal resolution, from March 2004 to April 2005. Measurements were collected in an urban area, where was characterized by heavy car traffic. Multiple sensors measured concentrations of CO (**μ**g/m 3), non-metanic hydrocarbons (NMHC)(**μ**g/m 3), NOx(ppb), and the temperature and relative humidity. At the measure site, conventional analyzers were used to provide ground truth of the target pollutants. Then, multisensor device is used to jointly measure all pollutants. In summary, observed CO concentration is much higher than NOx and NMHC (Figure 1 left panel); and the joint distribution of density between CO concentration and NMHC/NOx indicated strong correlations (Figure 1 right panel).

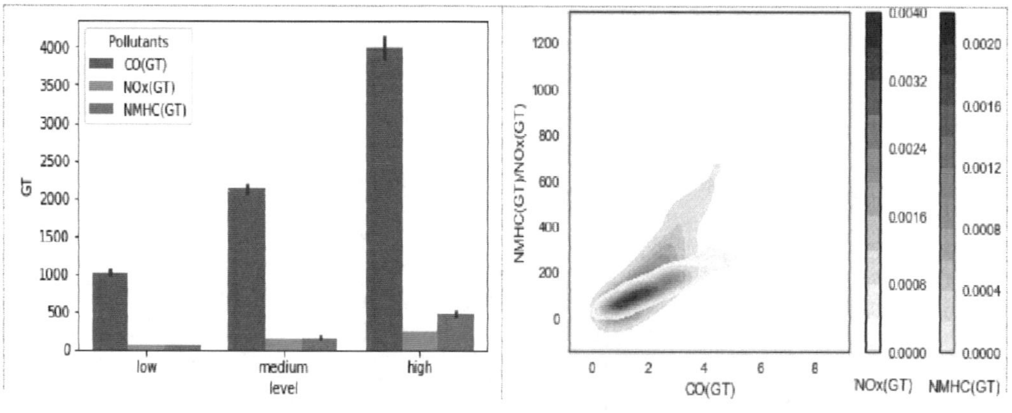

Figure 1. Left panel is the distribution of Pollutants concentrations on three levels (low,high,medium; Blue bar= CO, yellow bar= NOx, red bar= NMHC

Right panel showed correlation between CO concentration and NMHC/NOx concentrations.

We employed an Artificial Neural Network (ANN) to model the concentration dynamics of CO, NOx, and NMHC. ANN is a fully connected neural network (Figure 2), in which each neuron calculate a linear combination of previous layer neurons' output and embed with a non-linear activation function:

$$node = \sigma(\sum_{i=1} x_i \cdot w_i) \quad (1)$$

where σ is sigmoid activation function ($\phi(x) = \frac{1}{(1+e^{-x})}$), x_i and w_i represent values from previous layers' node and corresponding weighting parameters.

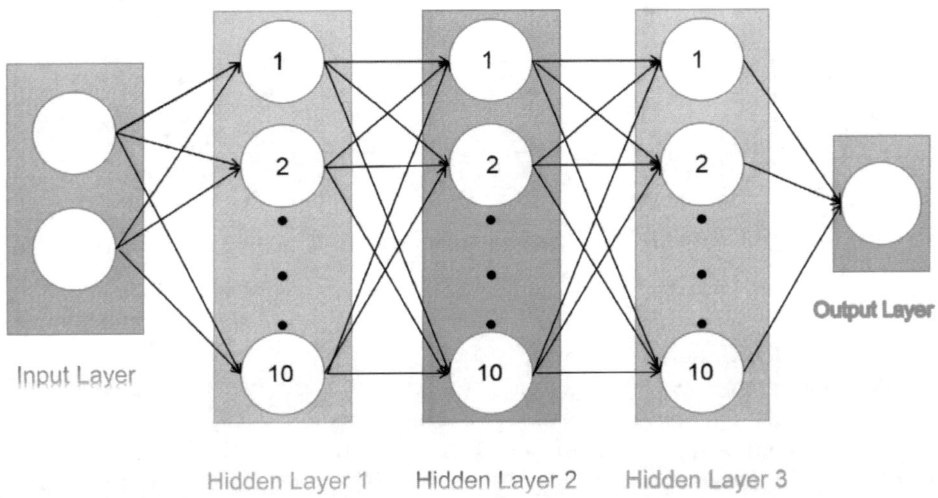

Figure 2. ANN Model structure. A 4-layer neural network with two inputs, three hidden layers of 10 neurons each and one output layer.

The ANN structure is designed as input-h1-h2-h3-output. Temperature and relative humidity were used in input variable. h represents hidden layer. And the output variable is pollutant. We constructed four independent ANN models: ANN model 1 predicts CO concentration; ANN model 2 predicts NOx; ANN model 3 simulates NMHC dynamics; and the fourth ANN jointly predicts all pollutants together.. The experimental data were divided into training, cross-validation, and testing purposes. By gradient decent search method, the training process progressively find the minimum of mean square error(MSE). Our hypothesis is that pollutant concentration prediction could benefit from joint prediction across multiple targets.

3. Results and discussion

3.1.Model prediction of single pollutant

All ANN models were set up with the same configuration, using 3 hidden layer, 10 neurons each layer, number of epoch was 150, and batch size was 1000. To avoid overfitting, the data set was randomly split into two sets, 80% for model training (to compute the gradient and updating of the network parameters, such as weights and biases—the training set) and 20% for model testing. The type of hidden (and output) unit activation function is relu function, and the type of optimization algorithms is Adam Optimization Algorithm [7].

Individual modeling of CO, NOx, and NHMC with ANN is not satisfactory in this case. CO, NOx, and NHMC only models had relatively large MSE of 0.027, 0.025, and 0.030, respectively. Moreover, the correlation between observed and predicted CO, NOx, and NHMC were smaller than 0.5 in general, indicating potential model deficiency (Figure 3).

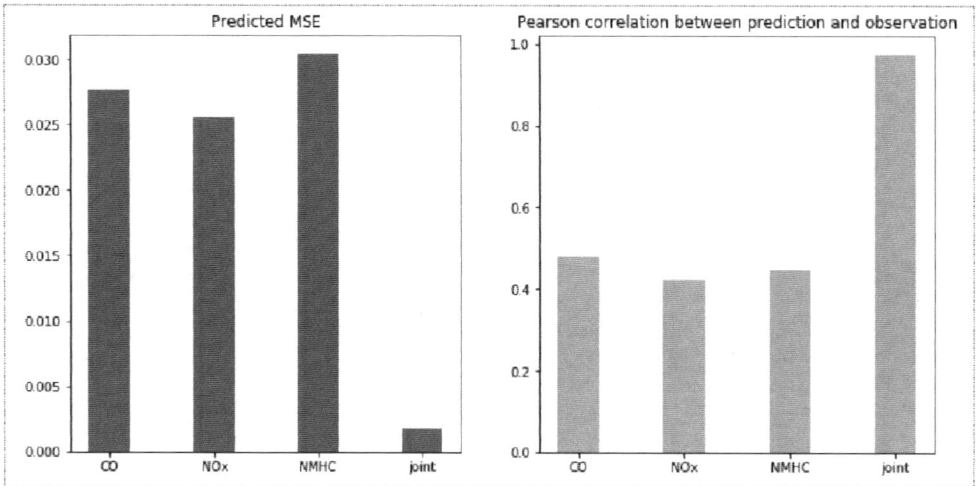

Figure 3. left panel is Predicted MSE for ANN Models of CO,NOx, NMHC and joint prediction. Right panel is the Pearson correlation between prediction and observation.

3.2.Joint prediction of multiple pollutants

Using the same model configuration as well as the same model complexity, we also jointly trained and predicted CO, NOx, and NMHC dynamics altogether. We found that the joint prediction could lower the MSE down to 0.00018 and enhance the model- data correlation to 0.97 (Figure 3). It supported our hypothesis that CO, NOx, NMHC in urban area were generated form the same source (e.g., vehicle emissions) could be jointly simulated. And more importantly, although three different chemical species, their time series encode similar information thus could benefit ANN modeling when the model enable the prediction of any species through learning from dynamics from other species.

3.3.limitation and future work

In this study, we limited the pollutants of interest to be CO,NOx and NMHC, and used them to train ANN models. However , there are other pollutants could also be potentially useful to help the joint prediction of CO, NOx, and NMHC dynamics. Furthermore, ANN model performance is highly dependent on parameters tuning. We did not go through all possibilities of parameters, such as the total possible value of nodes in hidden layers to determine the structure of ANN model. Appropriate network structure is achieved through experience and many times of trial and error, which require a large diversity of training for operation. In addition, we also acknowledged that we used a dataset from only one signal site, which is one of the main street in the centre of an Italian city. The spatial and temporal characteristics are limited based on just one site for one year, which may not present overview of the variations in the pollution situation.

In the future, we will focus on more pollutants at different sites to improve the accuracy of prediction and generalize our ANN modeling of multiple pollutant dynamics. In addition,to achieving appropriate network structure is to list as more possibilities as possible and to use trail and error way to find the good parameters.

4. Conclusions

Monitoring and modeling the temporal evolution of multiple primary pollutants over urban area is critically important to better understand their source-sink dynamics.

However, most of the available observations were made for each individual pollutant at different locations, in which case the quality of data is highly sensitive to instrumental malfunction. In this study, we used advanced machine learning technique to demonstrate that joint measurements of multiple primary pollutants not only benefit the spatial consistency of different pollutant measurements, but also favored the modeling effort if joint modeling of multiple pollutants was employed. We showed that by jointly modeling CO, NOx and NMHC, machine-learning model could learn information of one pollutant from the time series dynamics of another pollutant and improve the overall modeling accuracy.

5. References

[1] Levy, R. J. (2015). Carbon monoxide pollution and neurodevelopment: a public health concern. *Neurotoxicology and teratology, 49, 31-40.*

[2] Capşa, D., Barsan, N., Felegeanu, D., Stanila, M., Joita, I., Rotaru, M., & Ureche, C. (2016). Influence of climatic factors on the pollution with nitrogen oxides (NOx) in city, Romania. *Environmental Engineering & Management Journal (EEMJ), 15(3).*

[3] De Vito, S., Piga, M., Martinotto, L., and Di Francia, G. (2009). CO, NO2 and NOx urban pollution monitoring with on-field calibrated electronic nose by automatic Bayesian regularization. *Sensors and Actuators B: Chemical, 143(1), 182-191.*

[4] Rappenglück, B., & Fabian, P. (1999). Nonmethane hydrocarbons (NMHC) in the greater Munich area/Germany. Atmospheric Environment, 33(23),3843-3857.

[5] Shaban, K. B., Kadri, A., & Rezk, E. (2016). Urban air pollution monitoring system with forecasting models. *IEEE Sensors Journal, 16(8), 2598-2606.*

[6] Arunachalam, S., Valencia, A., Akita, Y., Serre, M. L., Omary, M., Garcia, V.,&Isakov, V. (2014). A method for estimating urban background concentrations in support of hybrid air pollution modeling for environmental health studies. *International journal of environmental research and public health, 11(10), 10518-10536.*

[7] Kingma, D. P., & Ba, J. (2014). Adam: A method for stochastic optimization. *arXiv preprint arXiv:1412.6980.*

IOP Conf. Series: Earth and Environmental Science **356** (2019) 012011 doi:10.1088/1755-1315/356/1/012011

Distribution of Soil Erodibility K Value in Gansu Province Based on Positioning Monitoring Data

YANG Ye[1,2], ZHANG Song-lin*[1] and WANG Lei[2]

1.Institute of Geography and Environmental Sciences, Northwest Normal University, Lanzhou, China.
2.School of Environmental Science and Engineering, Xiamen University of Technology, Xiamen, China.
E-mail: 15339135388@163.com

Abstract. This paper explored the spatial distribution characteristics of soil erodibility K value in Gansu Province by using the calculation model of soil erodibility factor. The influencing factors of the K value was analysed by traditional statistics and geostatistics. The located monitoring data of the natural ecological station was supported by China Ecological Research Network. The following conclusions could be obtained according to this study: (1) The average soil erodibility K of Gansu Province was 0.331 t·hm^2·h/ (hm^2·MJ·mm). (2) Gansu Province were mainly medium-high and high erosion areas. (3) The organic matter, rainfall, slope and elevation could influence K value. The rainfall and organic matter content were negatively correlated to the K value. And the slope and elevation have the law of inverted U-shaped distribution with the K value.

1.Introduction

The soil erosion areas account for 49.65% in Gansu Province according to the results of the first national water census [1]. The ecosystem of Gansu Province is severely constrained due to the soil erosion. Studying and fixing this problem could strongly promote sustainable development of agro-ecology.

The soil erodibility refers to the possibility of the soil being destroyed which is usually measured as K value [2]. The K value can be calculated as the amount of soil erosion caused by the unit rainfall erosivity on the standard area. The research methods of soil erodibility mainly include chemical properties determination method, instrument measurement method and mathematical model method, etc. The mathematical model of Erosion-Productivity Impact Calculator (EPIC) was widely used due to its advantages for the large areas.

The positioning monitoring data was supported by the China Ecological Research Network. The map of soil erosion in Gansu was drawn by ArcGIS for the first time which was used to research the related characteristics, influencing factors and spatial distribution of soil erodibility factors.

2.Materials and methods

In this study, the data is derived from the positioning monitoring data of China Ecology Research Network from 1998 to 2010. The EPIC model[3] was used to calculate the K value. The formula is:

$$K=\left\{0.2+0.3\exp\left[-0.0256\times S_a\left(1-\frac{S_i}{100}\right)\right]\right\}\times\left(\frac{S_i}{C_l+S_i}\right)^{0.3}\left[1-\frac{0.25C}{C+\exp(3.72-2.95C)}\right]\times\left[1-\frac{0.7S_n}{S_n+\exp(-5.51+22.9S_n)}\right] (1)$$

$$S_n = 1 - S_a/100 \quad (2)$$

Content from this work may be used under the terms of the Creative Commons Attribution 3.0 licence. Any further distribution of this work must maintain attribution to the author(s) and the title of the work, journal citation and DOI.
Published under licence by IOP Publishing Ltd

Where S_a is the content of sand (%), S_i is the content of silt (%), C_l is the content of clay, C is the content of organic carbon (%). However, the system of unit is different between data and the EPIC model. The soil data was transformed based on the Hermite mathematics [4]. The soil data was transformed by the software of MATLAB R2014a. The accuracy of calculation results was verified by the SPSS 22.0. The GS+ 7.0 calculated the optimal semi variogram theoretical model [5]. The ArcGIS 10.2 geographic software drawn the spatial distribution map of the soil erodibility K value.

3.Results and discussion.

3.1 The soil properties

The researched soil includes 7 classes 21 types, 47 subtypes, 64 genera and 106 species in this study. (Table 1).

Table 1. The types and areas of soil in Gansu Province

Classes	Type	Area (km^2)	Classes	Type	Area (km^2)
Semi-leaching soil	Cinnamon soil	16497.58	Arid soil	Gray earth	15767.88
	Black soil	1394.03	Leaching soil	Dark brown soil	4361.51
Early breeding soil	Gray cinnamon soil	6732.70		Yellow cinnamon soil	21.34
	Chestnut soil	2878.11		Yellow brown soil	977.16
	Windy sand	24030.68		Brown soil	
	Red clay	3532.43		Gray desert soil	7276.30
	Yellow loess	36028.67		Gray brown desert soil	2751.38
	New fill	648.99			7605.80
		3667.17	Desert soil	Brown desert soil	
		9223.28		Irrigation desert soil	46009.66
Calcareous soil	Black earth	9652.16		Irrigation	6710.69
	Black earth(II)				
	Chestnut soil		Artificial soil		1747.54

The Coefficient of Variation (CV) was described the range of variation. The pH value and organic matter content of soil is shown the Table 2.

Table 2 Soil organic matter and pH statistics in Gansu Province

	Max	Min	Average	Standard deviation	Range	CV (%)
Organic Matter (g/kg)	122.70	1.50	23.42	22.60	121.20	96.50
pH	9.30	5.80	8.07	0.64	3.50	7.90

3.2 Calculating data

The MATLAB R2014a software was used to transform date. The coefficient R^2 and Pearson correlation coefficient were verified the converted result. The results below can be obtained from Figure 1: R^2=0.943, Pearson correlation coefficient =0.972.

IOP Conf. Series: Earth and Environmental Science **356** (2019) 012011 doi:10.1088/1755-1315/356/1/012011

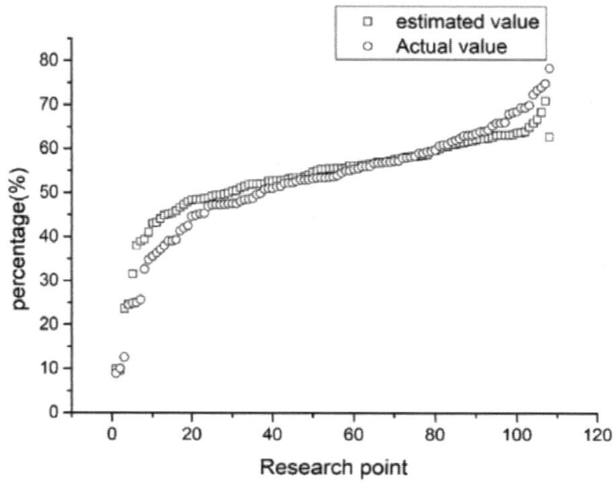

Figure 1. Interpolated estimates and measured values

The K value was calculated by the EPIC model. The range of the soil erodibility K value is 0.068~0.417t·hm²·h/(hm²·MJ·mm). The average value is 0.331t·hm²·h/(hm²·MJ·mm). The result is shown the Table 3:

Table 3 The erodibility K Values of the Gansu province

Classes	Max	Min	Average	Standard deviation	Range	CV (%)
Early breeding soil	0.397	0.067	0.329	0.103	0.330	31.268
Semi-leaching soil	0.392	0.235	0.307	0.040	0.156	13.137
Calcareous soil	0.417	0.281	0.339	0.039	0.136	11.560
Arid soil	0.380	0.257	0.335	0.039	0.123	11.782
Leaching soil	0.399	0.191	0.317	0.060	0.207	18.723
Desert soil	0.393	0.272	0.349	0.042	0.122	12.141
Artificial soil	0.413	0.304	0.351	0.027	0.108	7.781
Total	0.417	0.067	0.331	0.057	0.349	17.379

The degree of the soil erosion is divided into five categories: low corrosive soil ($K<0.15$), medium and low erodible soil ($0.15<K<0.25$), medium erodible soil ($0.25<K<0.30$), medium high Erodible soil ($0.30<K<0.35$) and highly corrosive soil ($K>0.35$).

3.3 Drawing the map

The GS+ analysis software was used to calculate the optimal semi-variogram models. The results shown that the Gaussian model is the most suitably. The Kriging interpolation method is used to drawn the spatial distribution of soil K values in Gansu Province (Figure 2).

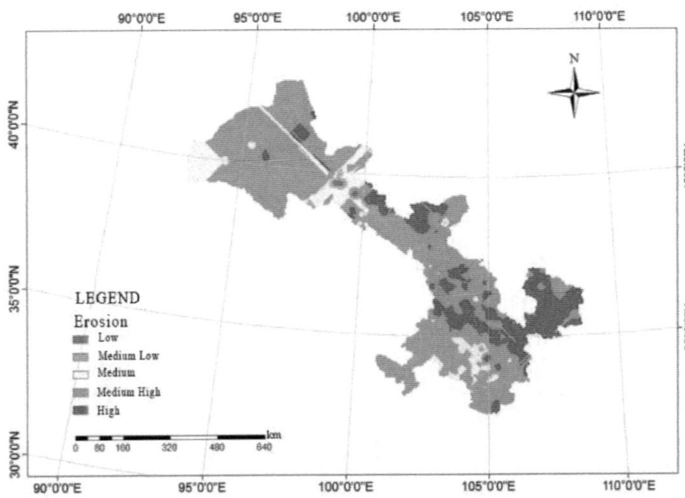

Figure 2. Spatial distribution of soil corrosion K-value in Gansu Province

Figure 2 shows the spatial distribution characteristics of the soil K values in Gansu Province. The major areas are the serious soil erosion areas. In summary, the high-erosion areas are mainly located in the Loess Plateau, the Tengger Desert and the vicinity of the Badain Jaran Desert. The low-erosion areas mainly distribute in the hinterland of the Hexi Corridor.

This paper contrasted that the K value of Gansu Province and other regions. Results show that: The K value of Gansu Province is higher than the Yimeng Mountain [6], the Tai Lake basin [7] and the Qinghai-Tibet Plateau [8], and slightly lower than Qinghai Lake Watershed [9].

3.4 Influencing factor

3.4.1 Organic matter

As shown in Figure 3: There is a medium negative correlation between the desert soil and the arid soil.

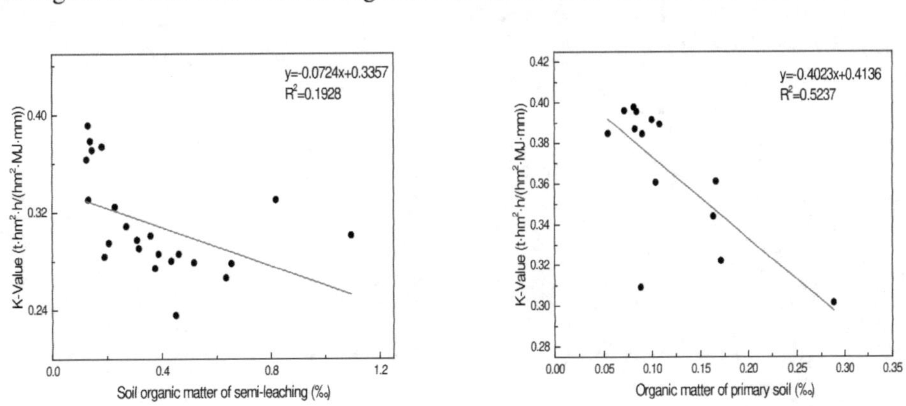

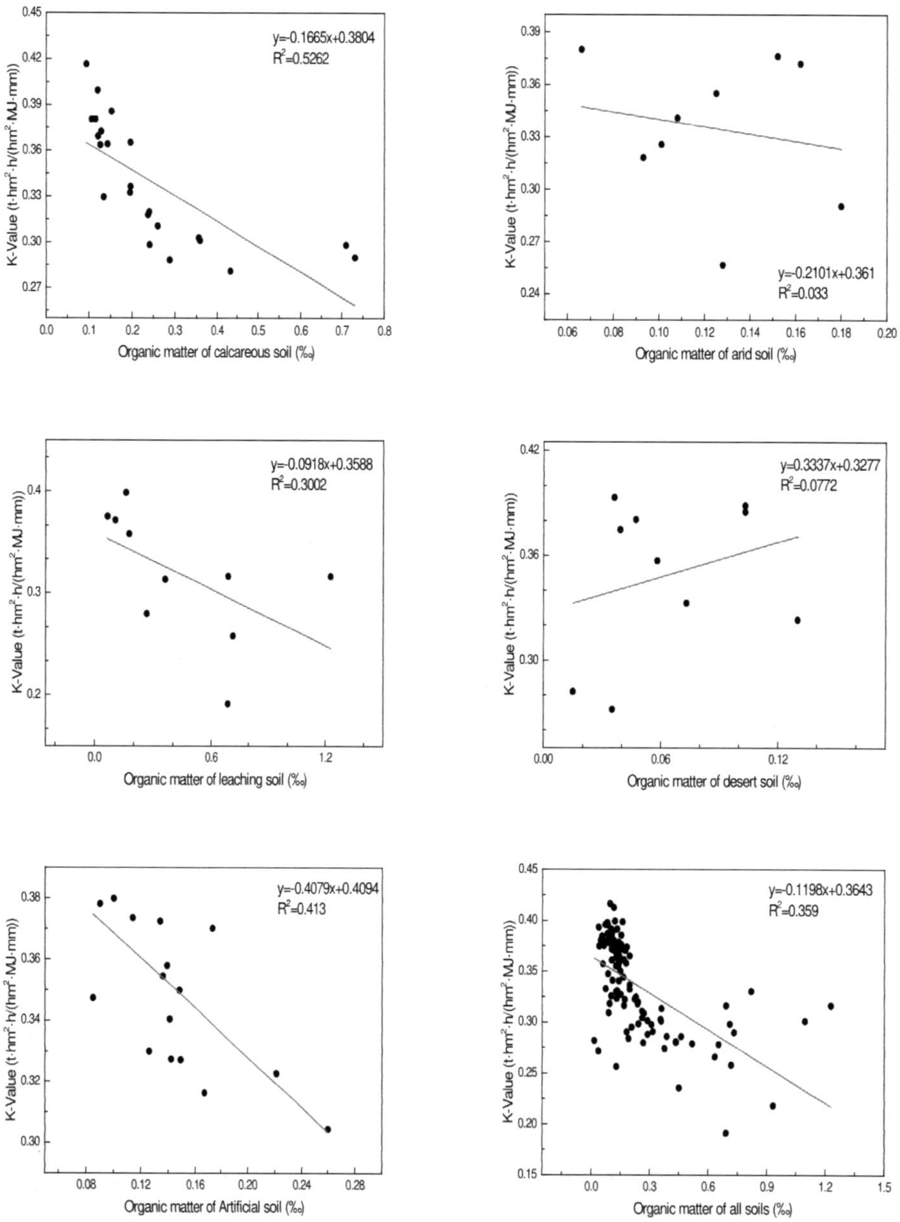

Figure 3. The Linear fitting diagram of soil type and organic matter content

The K value become diminish with the organic matter content increased[10]. Because of increasement of the soil agglomerate, the soil physical properties improve. However, the desert soil and arid soil have less organic matter content. So, the K value may affect by other factors.

3.4.2 The annual rainfall

The date of annual rainfall is divided into eight intervals (the annual rainfall <100 mm, 100~200 mm, 200~300 mm, 300~400 mm, 400~500 mm, 500~600 mm, 600~700mm and >700 mm) for conveniently analyze data. Figure 4 shows that: when the annual rainfall is less than 500 mm [11], there is the positive correlation for the K value and precipitation. When the annual rainfall is more than 500 mm, there is no significant correlation between the K value and precipitation.

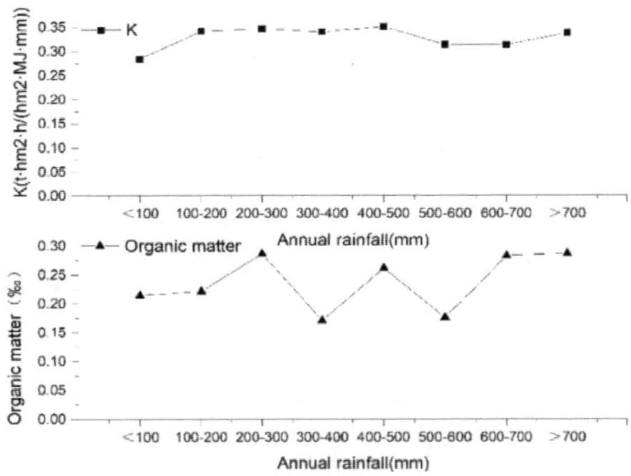

Figure 4. Distribution of annual rainfall and organic matter

The results show that in the interval of the 500-600mm. The value of soil average physical sand is larger and the physical clay value is smaller. This situation will diminish the K value. In the interval that the annual rainfall exceeds 600mm, there is the high organic matter content. It's may be the reason for the transformation law of K value is different from another intervals.

3.4.3 Topographical factors

In this paper, the slope and elevation data extracted from the Gansu Province Digital Elevation Model (DEM) map for researched topographical factors. As is shown the Figure 5 and Figure 6: the slope and elevation have the law of inverted U-shaped distribution with the K value. Besides, there is the specific interval of slope and elevation. the K value is the maximum. K value is gradually decreasing on both sides of this interval.

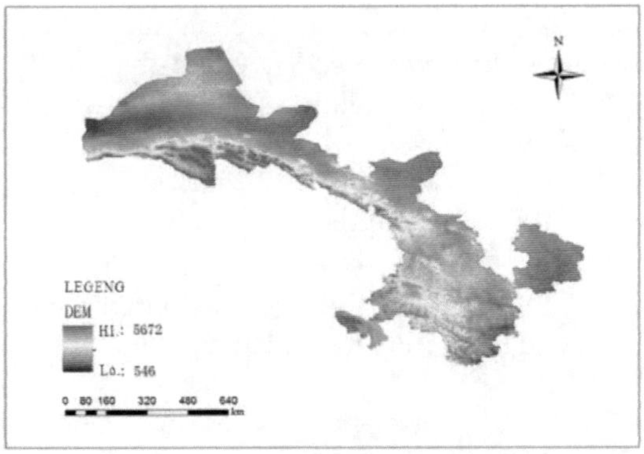

Figure 5. The elevation distribution map of Gansu province

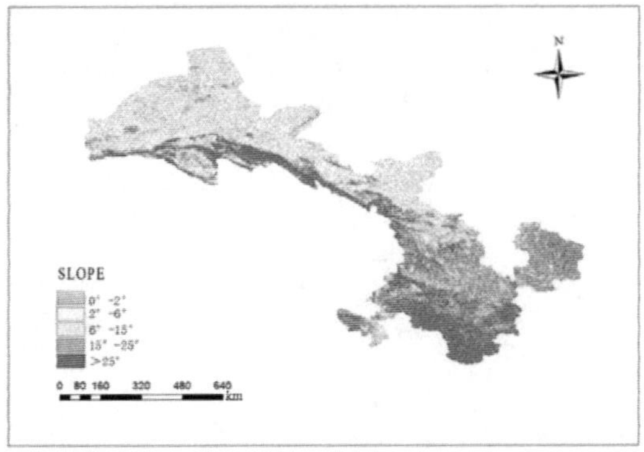

Figure 6. The slope distribution map of Gansu province

4. Conclusion

1. The EPIC model can accurately count the K value of Gansu province.
2. The soil of Gansu Province is weakly alkaline, and the organic matter content is greatly discrepant across the province.
3. The great mass of area is moderately high and high erosion areas on Gansu Province. The soil loss risk is higher other area.
4. The soil erodibility K value was affected by factors, such as rainfall, organic matter content, slope and elevation. The rainfall and organic matter content are negatively correlated with the K value. the slope and elevation have the law of inverted U-shaped distribution with the K value.

5. References

[1] Yao J Z 2015 Thoughts on Soil and Water Conservation Work in Gansu *Province Soil and Water Conservation in China* (Gansu: Gansu Provincial Water Resources Department) chapter 3 pp 1-2, 11.
[2] Wang B and Zheng F L 2015 Soil Erodibility for Water Erosion: A Review *Research of Soil and Water Conservation* (Yangling: Northwest A&F University) chapter 20 pp 277-286.
[3] WILLIAMS J R. 1983 EPIC- a New Method for assessing erosion' s effect on soil *productivity Journal of Soil and Water Conservation*, chapter 38 pp 381 - 383.
[4] Yan R and Yan S J 2013 The Accuracy Verification of the Different Soil Texture Transformation Models in Cha-kou Watershed *Journal of Shanxi Agricultural University* (Taigu: Shanxi Agricultural University) chapter 33 pp 109-113.
[5] Shi Z and Li Y 2006 *Application of geostatistics in soil science* Beijing (China Agriculture Press)
[6] Jin S S and Zhang R H 2017 Spatial Variability on K Value of Soil Erodibility in Typical County of Yimeng Mountain Area *Chinese Journal of Soil Science* (Taian: Sandong Agricultural University) chapter 48 pp 278-284.
[7] Bu Z L and Yang L Z 2002 Soil erodibility (K) value and its application in Taihu lake catchment *Acta Pedologica Sinica* (Nanjing: Institute of soil science) chapter 39 pp 296-300.
[8] Liu B T and Tao H P 2014 Spatial distribution characteristics of soil erodibility K Value in Qinghai-Tibet Plateau *Bulletin of Soil and Water Conservation* (Chengdu: Institute of Mountain Hazards and Environment) chapter 34 pp 11-16.
[9] Liu J F and Li S J 2006 Soil erodiable K in the catchment of Qinghai lake *Arid Land Geography* （Nanjing: Nanjing Institute Of Geography & Limnology CAS） chapter 29 pp 321-326.
[10] Yang C M 2008 Quantitative Study on the Relation of Soil Organic Content and Soil and Water Losses Research of Soil and Water Conservation (Taiyuan: Institute of Soil and Water convention of Shanxi) chapter 15 pp 177-179.
[11] Sun D F and Wang L 2007 Effects of farmlands' soil erosion under different intensities of rainfall in loess hilly regions Research of Soil and Water Conservation (Lanzhou: Gansu Agricultural University) chapter 14 pp 16-18.

Acknowledgments

The National Natural Science Foundation of China (51068025), Gansu Natural Science Foundation Project (1308RJZA302), Gansu Provincial Key Laboratory of Oil and Gas Resources Research Open Fund (SZDKFJJ20150606) and Gansu Provincial Key Science Ecological Science Funding.

Terrain numerical simulation based on RANS/LES hybrid turbulence model

Lu Xiegui[1], Chen Qiuhua[2], Qian Changzhao[2], Chen Changping[1,2,3]
1 School of Architecture and Civil Engineering, Xiamen University, Xiamen, China;
2 Haixi Wind Engineering Research Center, Xiamen University of Technology, Xiamen, China;
3 Xiamen Ocean Vocational and Technical College, Xiamen, China;
E-mail: cpchen@126.com

Abstract. Based on the computational fluid dynamics (CFD) method, the terrain of 2 km near the proposed humanoid landscape bridge in Xiamen is taken as the research object. Firstly, the three-dimensional topographic map is established. Because the selected terrain is relatively flat, the bridge area is adopted in the form of spline curve. The boundary is extended, and the exponential rate of wind speed boundary conditions is applied to the boundary to avoid non-0m/s wind speed from the inflow near wall boundary. The RANS/LES hybrid mode is used to numerically analyze the external flow field of the terrain. The κ - ε - v^2 turbulence model of the three equations is used as the RANS part of the RANS/LES hybrid turbulence model. According to the Boussinesq hypothesis and dimension-analysis, the differential form equations of v^2 are rationally simplified, and the algebraic form of the normal stress equation of v^2 and the new turbulent viscosity equation are obtained and written into FLUENT in UDF form. By comparing and analyzing the standard κ - ε turbulence model and the $SST\kappa$ - ω turbulence model, the wind speed amplification factor and the variation law of wind parameters along the height are studied, and the canyon effect and the reduction effect of the wind field are discussed. The results show that under the same calculation conditions, the mixed turbulence model can accurately reflect the change of airflow in the near-wall wall, and more accurately capture the wind field away from the bridge near the wall.

1. Introduction

In many numerical simulations of terrain, there are often engineering backgrounds of long-span bridges. This makes our focus not only on accurately simulating the near-surface flow characteristics and capturing the flow separation characteristics, but also on the flow field far from the bridge location in a certain height near the ground. Therefore, we attach great importance to the selection of turbulence model. Li Yongle et al[1-2] used the laminar flow model to simulate the 8km×8km regional terrain wind field in the deep-cut canyon bridge site, and compared the wind speed along the height under different working conditions, and proposed the general classification of the canyon wind field. According to the variation characteristics of the wind speed along the main beam, the joint distribution of average wind speed and wind angle of attack is obtained. Zhang Yue et al[3] studied the wind environment near the site of a cable-stayed bridge in the western mountainous area, and compared the actual measurement with the numerical simulation using the $SST\kappa$ - ω turbulence model and the Realizableκ - ε turbulence model. Zhang Xiangxu et al[4] analyzed the wind field of a large-span arch bridge in the mountainous area of western China, and used the circular curve and cosine curve to expand the terrain boundary. The $SST\kappa$ - ω turbulence model was used to simulate the wind field.

Content from this work may be used under the terms of the Creative Commons Attribution 3.0 licence. Any further distribution of this work must maintain attribution to the author(s) and the title of the work, journal citation and DOI.
Published under licence by IOP Publishing Ltd

Uchida and Ohya[5] used the large eddy simulation method to study the air flow on complex terrain, and discussed two different grid schemes, and obtained the distribution of average wind speed and pulsating wind speed. In the turbulence model, large eddy simulation is a turbulent numerical simulation method between direct numerical simulation (DNS) and RANS[6], which is the most rapidly developing turbulence model in recent years. The main idea is to directly solve the large-scale eddy in the flow field and establish a model for small-scale turbulent pulsation. For complex terrain where the flow field changes more severely, the resolution requirements of the grid are also higher. In order to solve the above-mentioned problem: the near-wall wall capture flow characteristics, the bridge site is concerned with the wind field distribution. This paper adopts the RANS/LES hybrid mode, and the idea is to simplify the use of Reynolds at the near wall to reduce The number of small grids, while large eddy simulation in the far field, captures the flow field characteristics at the bridge site height. Among them, the three-equation κ - ε - v^2 turbulence model is chosen as the RANS model in the mixed turbulence model.

2. κ - ε - v^2 Turbulence model simplification

LAUNDER[7] pointed out that the normal stress of the vertical wall has a great influence on the turbulent kinetic energy and accounts for a large proportion of the eddy viscosity coefficient when studying the low Reynolds number turbulent flow near the wall. In response to this conclusion, Durbin[8] constructed a vertical wall normal stress v^2 equation based on the standard κ - ε turbulence model, and closed the RANS equation by using the near-wall elliptic relaxation function equation (f) to obtain the κ - ε - v^2 turbulence model. And the governing equations for v^2 and f are [8]:

$$\frac{\partial(\rho v^2)}{\partial t} = \frac{\partial}{\partial x_i}\left[\left(\mu + \frac{\mu_t}{\sigma_k}\right)\frac{\partial v^2}{\partial x_i}\right] + \rho\kappa f - \rho v^2 \frac{\varepsilon}{\kappa}$$

$$L^2 \frac{\partial}{\partial x_i}\left(\rho \frac{\partial f}{\partial x_i}\right) = \rho f + \rho c_1 \frac{(2/3 - v^2/\kappa)}{T} - c_2 \frac{P_\kappa}{\kappa}$$

In this equation,

$$\begin{cases} \mu_t = \rho c_\mu v^2 T \\ P_\kappa = \mu_t \dfrac{\partial u_j}{\partial x_i}\left(\dfrac{\partial u_i}{\partial x_j} + \dfrac{\partial u_j}{\partial x_i}\right) \\ T = \max\left(\dfrac{\kappa}{\varepsilon}, 6\left(\dfrac{v}{\varepsilon}\right)^{0.5}\right) \\ L = 0.3\max\left(\dfrac{\kappa^{1.5}}{\varepsilon}, c_\eta\left(\dfrac{v^3}{\varepsilon}\right)^{0.25}\right) \end{cases}$$

And $c_\mu = 0.19$, $c_\eta = 70$, $\sigma_\kappa = 1.0$, $c_1 = 1.4$, $c_2 = 0.3$

The κ - ε - v^2 turbulence model does not use the wall function in the near wall area, and the turbulence has anisotropic characteristics, which can more accurately simulate the flow separation phenomenon.

To build a RANS/LES hybrid turbulence model, the κ - ε - v^2 turbulence model needs to be simplified. According to the Boussinesq hypothesis[9], the momentum transfer caused by turbulent eddies can be expressed by the eddy viscosity coefficient:

$$\tau_{ij} = 2\mu_t S_{ij}^* - \frac{2}{3}\rho k \delta_{ij}$$

Here μ_t is called the eddy viscosity coefficient. Let $i = j = 2$, and substitute $v_t = c_\mu v^2 T$ in the κ - ε - v^2 turbulence model into the eddy viscosity coefficient to get:

$$\frac{v^2}{\kappa} = C_1\left(1 - C_2 \frac{v^2}{\varepsilon}\frac{\partial v}{\partial y}\right)$$

In this equation, C_1 and C_2 are constants. Through dimensional analysis, taking $y^* = y c_\mu^{1/4} \kappa^{1/2} / v$, and $l^* = \kappa^{3/2}(\varepsilon y)$, the equation can be rewritten as a dimension form[10]:

$$\frac{v^2}{\kappa} = C_1\left(1 - g\left(y^*, l^*\right)\right)$$

Where $g\left(y^*, l^*\right)$ is the pending function. Assuming that the value tends to infinity, ie away from the wall, the turbulence should be a uniform flow, which exists[10]:

$$\frac{v^2}{\kappa} = C_1\left(1 - g\left(y^*, l^*\right)\right) \xrightarrow{y^* \to \infty} C_1$$

When y^* is very infinitely close to 0, the normal kinetic energy tends to 0, and you can get:

$$\frac{v^2}{\kappa} = C_1\left(1 - g\left(y^*, l^*\right)\right) \xrightarrow{y^* \to 0} 0$$

It is assumed that the wall damping effect is only related to the geometry, and the influence of the turbulent vortex scale l^* can be ignored. Using the exponential function instead of $g\left(y^*, l^*\right)$, the algebraic form of the normal stress v^2 equation can be obtained:

$$\frac{v^2}{\kappa} = A\left(1 - \exp\left(-\frac{y^*}{B}\right)\right)$$

Where A and B are pending coefficients. The κ - ε - v^2 turbulent viscosity in the final turbulence model translates into:

$$v_t = c_\mu A\left(1 - \exp\left(-\frac{y^*}{B}\right)\right)\kappa T$$

In order to determine the undetermined coefficients A and B, the original κ - ε - v^2 turbulence model can be activated by using the TUI command in FLUENT to obtain the velocity variances v^2, y^* and the turbulent kinetic energy value κ at different monitoring points, and curve fitting is performed, thereby obtaining $A = 0.638$ and $B = 69.71$ in reverse.

Write a *UDF* (User Define Function) to write the new turbulent viscosity to the definition of the turbulent viscosity of the RANS/LES hybrid model.

3. CFD numerical simulation model

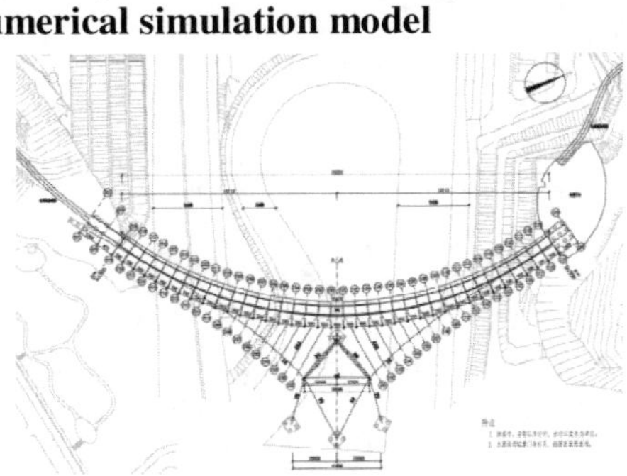

Figure 1. Bridge floor plan

Figure 2. Bridge renderings

3.1. Geometric Modeling and Meshing

This paper takes the site selection of a proposed pedestrian walkway in Xiamen as the research object. The long-span humanoid bridge has high flexibility, is more sensitive to incoming wind, and is more prone to bridge flutter and vortex vibration.Therefore, it is of certain importance to study the wind environment of bridges in this area, as shown in Figure 1, Figure 2.

In this paper, we use *GoogleEarth* combined *SketchUp* to obtain $2km*2.5km$ geographic data of the bridge site and establish a three-dimensional geometric model. The location area of the bridge is located in the urban area, and the overall terrain is relatively flat. In order to avoid the assumption that the near-surface boundary conditions do not satisfy the non-slip wall surface, the near-surface wall at the inlet is guaranteed to have a wind speed of $0\ m/s$. The spline is used to widen the terrain to a certain extent[4], so that the terrain data is at the same elevation at the boundary entrance, as shown in Figure 3.

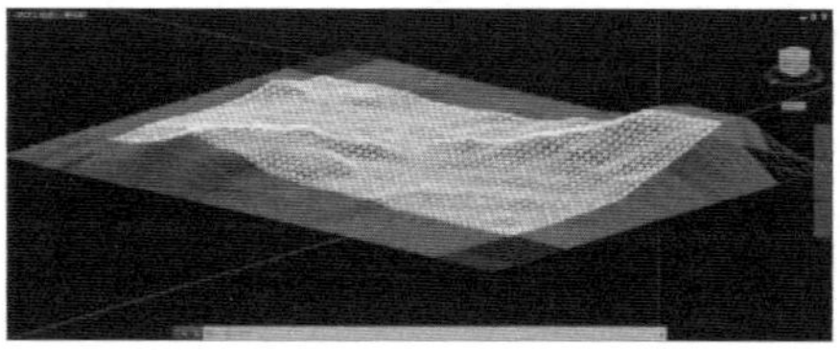

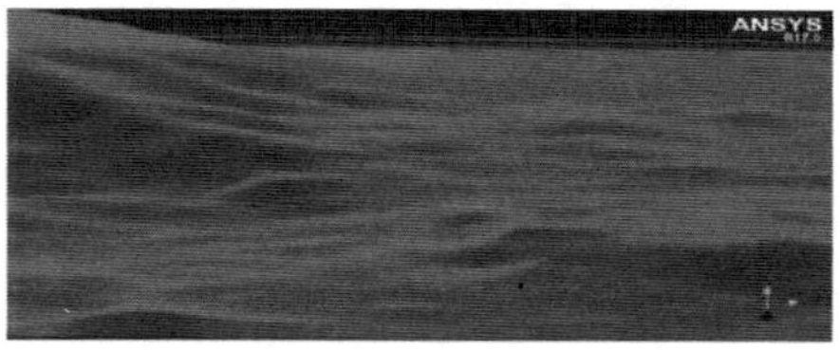

Figure 3. Bridge topography extension **Figure 4.** Structured grid generation

Use the *ICEM* in *Ansys* to divide the terrain into a three-dimensional structured grid, as shown in Figure 4. The maximum resolution of the grid is 4×4m, the height of the first layer grid is 0.3m, the expansion ratio between two consecutive cells was limited to 1.2, and finally 6.23 million hexahedral meshes are obtained.Grid quality is higher than 0.7.

3.2. Basic calculation of wind speed and wind direction

The basic wind speed is in accordance with the maximum wind speed in Xiamen City in one hundred years. The maximum wind speed at the height of 10 meters is 39.7m/s. According to the B type landform, the ground roughness coefficient is $\alpha = 0.16$, and the wind profile is distributed according to the exponential rate. The expression is as follows: $U(z) = 39.7 \times \left(\frac{Z}{10}\right)^{0.16}$, the inflow profile wind is realized by editing the UDF file. The wind speed at the speed entrance is as follows (where h_0 is the lowest point elevation of the wind speed entrance):

When elevation $H \geq h_0 + 350m$, $V_\infty = 70.12m/s$

When elevation $H \leq h_0$, $V_\infty = 0m/s$

When the elevation $h_0 \leq H \leq h_0 + 350m$, $V_\infty = 39.7 \times \left(\frac{Z}{10}\right)^{0.16}$

In this expression, Z is the point position elevation (unit: m) at the entrance boundary; V_∞ is the velocity inlet flow velocity (unit: m/s).

Statistical analysis of meteorological data in Xiamen for nearly 30 years, using the *OriginPro* software to draw a typical wind rose map of Xiamen, as shown in Figure 5. The figure shows the wind speed combined wind frequency distribution in 16 directions in Xiamen. It can be seen from the figure that the dominant wind direction in Xiamen is the E direction.

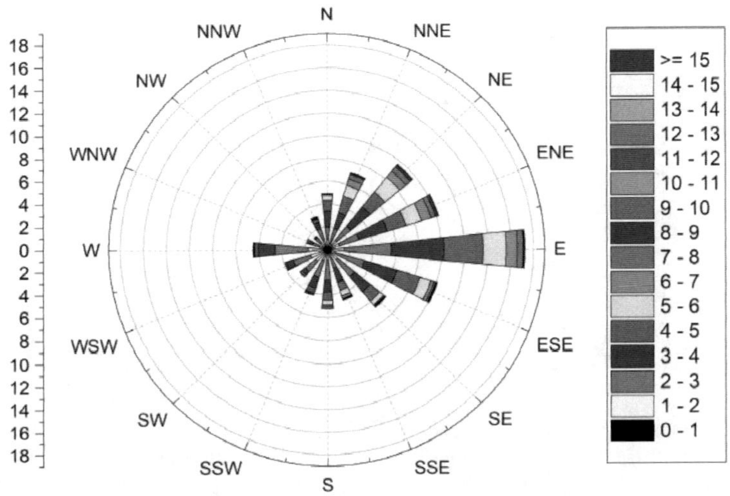

Figure 5. Distribution of wind speed combined wind frequency in Xiamen

3.3. Boundary conditions and turbulence models

The numerical simulation uses *Fluent* calculation software, which uses $SSTk-\omega$ model, standard κ - ε turbulence model and RANS/LES mixed turbulence model for comparison. Turbulence intensity is 1%, turbulent viscosity ratio is 2, time dispersion is second-order implicit. Gradient calculation by default $Least-Squares-Cell-Based$, the pressure interpolation method and other spatial discretes are second-order formats, and the *SIMPLE* algorithm deals with the coupling of pressure and velocity.

The boundary condition at the entrance is $Velocity-Inlet$, and the boundary condition at the exit is the $Pressure-Outlet$. The terrain is a non-slip wall boundary condition, and the surrounding and top surfaces are symmetric boundary conditions (*Symmetry*) . In the mixed turbulence model, the time step is 0.002 s. For this study the convergence criterion for all residuals was fixed at $10e^{-5}$.

3.4. Wind speed amplification factor

In the numerical simulation of terrain, the wind speed amplification factor under different wind direction conditions is usually used to judge whether the terrain has the effect of amplifying or reducing the wind speed for the incoming flow. It is generally considered that the ratio of the actual wind speed at the bridge location in the canyon to the wind speed at the inlet boundary condition at the same height is defined as the wind speed amplification factor[11]. The wind speed amplification factor is defined in the following equation:

$$C_u = \frac{v_1}{v_0}$$

In this expression: v_1 is the monitored wind speed at the bridge point, and v_0 is the inlet boundary wind speed at the same height.

4. Numerical simulation results analysis

4.1. Influence of three turbulence models on profile wind

Taking the E direction of the maximum probability wind direction as the inflow direction, the relative position of the bridge is selected as the west bridge tower, the bridge 1/4 span, the middle bridge span, the bridge 3/4 span and the east bridge tower as the monitoring points. The wind profile of three turbulence models and the wind speed amplification factor of mixed turbulence model at eight angles of attack are plotted, as shown in Fig. 6.

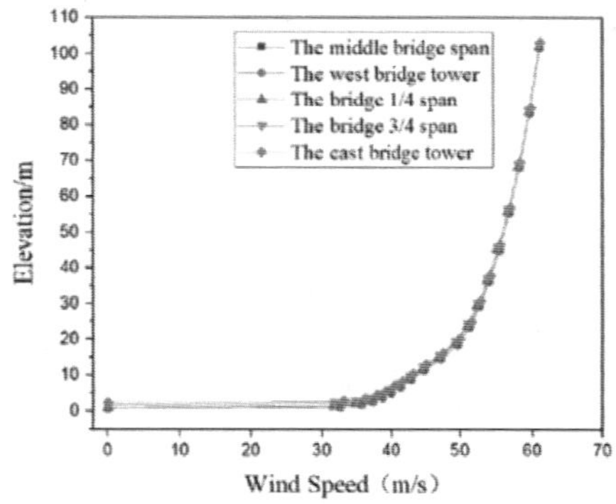

(a)Wind Profile at Midspan of Bridge under the Standard $\kappa - \varepsilon$ Turbulence Model

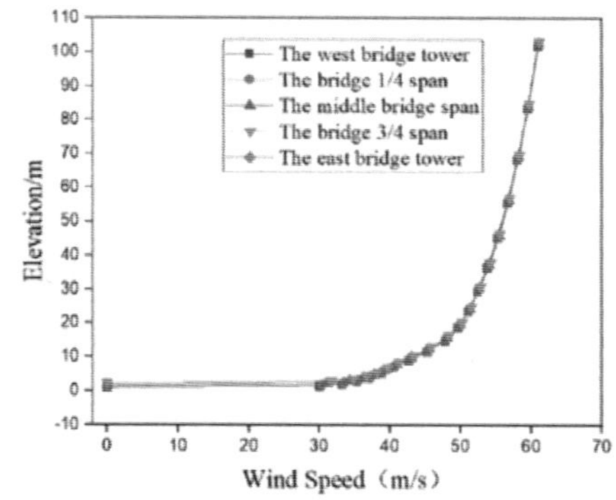

(b)Wind Profile at Midspan of Bridge under $SST\kappa - \omega$ Turbulence Model

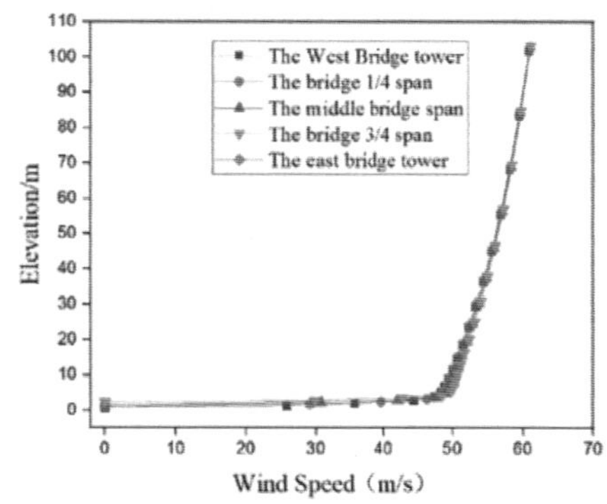

(c)Wind Profile at Midspan of Bridge under $\kappa - \varepsilon - v^2 / LES$ Turbulence Model

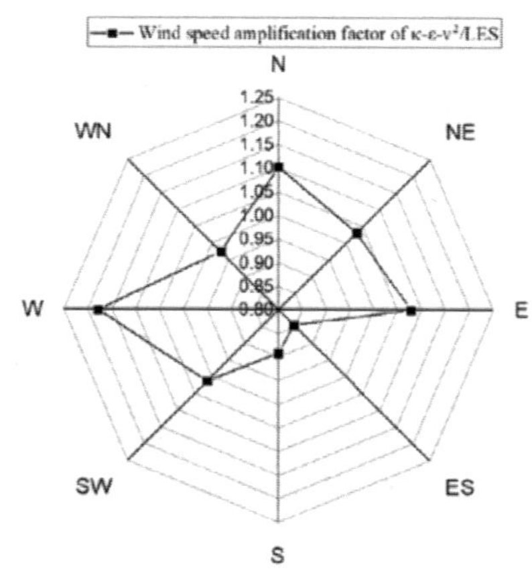

(d)wind speed amplification factor of $\kappa - \varepsilon - v^2 / LES$ turbulence model at eight angles of attack

Figure 6. Wind profile of three turbulence models and wind speed amplification factor of mixed turbulence model at eight angles of attack

As can be seen from the figures (a), (b), (c), there are some differences in the wind profiles presented by the three turbulence models. The wind speed of the $SST\kappa - \omega$ turbulence model varies smoothly along the elevation, while that of the $\kappa - \varepsilon - v^2 / LES$ turbulence model varies sharply along the elevation. From the figure (c), it can be seen that there is a wind speed close to 50 m/s in the 10 m elevation range, which is different from the other two turbulence models at the same altitude as the assumption of the inflow profile wind. The magnification effect of terrain on the wind speed is highly evaluated. It can be seen from figure (d) that the maximum wind speed amplification factor is 1.18 and the minimum reduced wind speed factor is 0.845 for the mixed turbulence model at eight angles of attack. It can also be seen from the figure that the annual wind direction E wind speed amplification factor is 1.08.

4.2. Velocity nephograms of three turbulence models at different contours

The height of 0.5m, 1m, 10m to 200m (10m interval) near the ground is taken as the observation object. We can see the velocity nephogram of the whole topographic field of three turbulence models in different contour ranges as shown in Fig. 7.

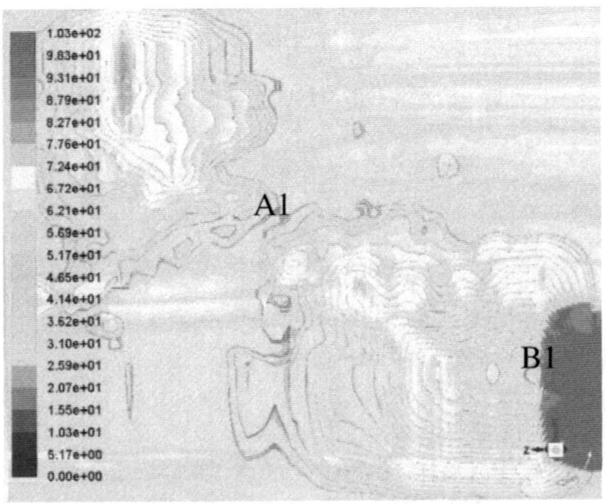

(a)Velocity nephogram of the standard κ - ε turbulence model at different contours

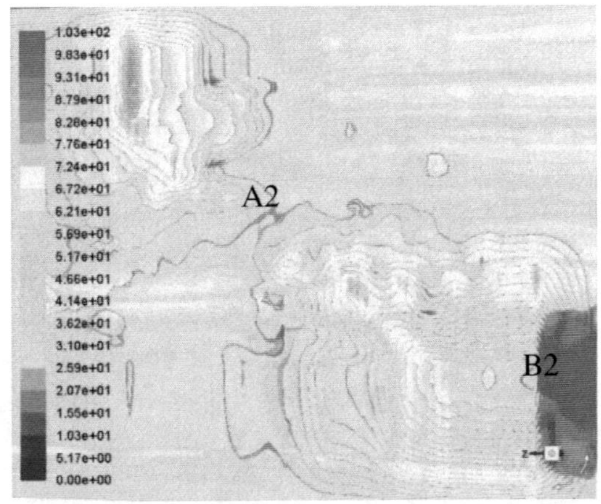

(b)Velocity nephogram of $SST\kappa$ - ω turbulence model at different contours

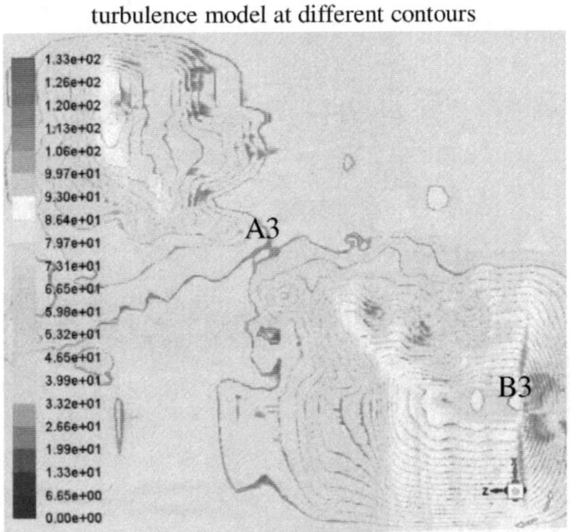

(c)Velocity nephogram of κ - ε - v^2 / LES turbulence model at different contours

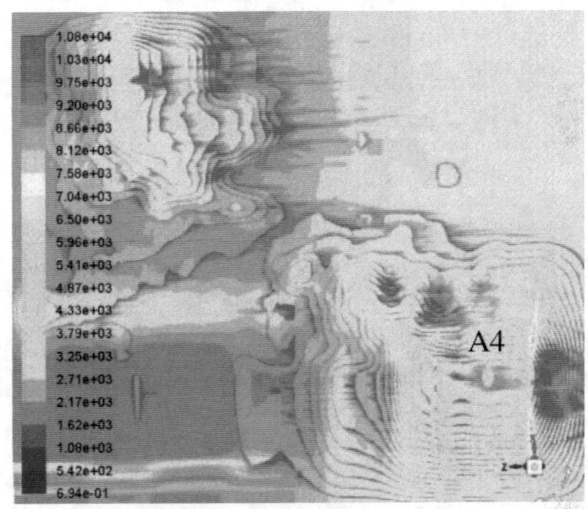

(d)Dynamic pressure cloud diagram of κ - ε - v^2 / LES turbulence model at different contours

Figure 7. Velocity nephograms of three turbulence models at different contours

From Fig. 7 (a), (b), (c), it can be seen that the three turbulence models are in a lower range of relative wind speed at the location of bridge, i.e. A1, A2 and A3 in the figure, and there is no Canyon effect of rapidly increasing wind speed. It can be seen from A4 in figure (d) that the dynamic pressure varies greatly and the wind speed distribution is not uniform at this position. By comparing B1, B2 and B3, it can be seen that the standard κ - ε turbulence model and $SST\kappa$ - ω turbulence model have poor wind capture characteristics at position B, and do not really reflect the real situation of wind speed change at that location. However, the κ - ε - v^2 / LES turbulence model can better reflect the real situation of wind field change at that location.

5. Conclusion

1) The numerical results show that compared with the other two turbulence models, the κ - ε - v^2 / LES mixing turbulence model is better in evaluating the wind speed in the wind profile. The reason is that the κ - ε - v^2 turbulence model used in the $RANS$ term of the mixing turbulence model ,which takes into account the anisotropic characteristics of the turbulence near the wall.

2) κ - ε - v^2 / LES mixed turbulence model performs well in numerical simulation of complex terrain, and can accurately reflect the actual appearance of terrain wind field. The reason is that the mixed turbulence model uses large eddy simulation at the far end of the wall. This simulation method is more accurate than other models in capturing the flow changes of space flow field.

3) It can be seen from the numerical simulation that there is no obvious Canyon effect in the location of the bridge under the condition of the topographic location. Under the condition of perennial wind direction, the wind speed amplification factor is 1.08, which is acceptable in practice.

6. References

[1] Li Yongle, Cai Xianyu, Tang Kang, et al. Numerical Simulation of Spatial Distribution Characteristics of Wind Field in Deep Canyon Area[J]. China Civil Engineering Journal, 2011(2): 116-122.
[2] LI Yongle, Tang Kang, CAI Xian-xi, et al. Composite wind speed standard for long-span bridges in deep canyon area[J]. Journal of Southwest Jiaotong University, 2010, 45(2).
[3] ZHANG Yue. Study on wind field characteristics of cable-stayed bridge in western mountainous terrain[J]. Journal of Wuhan University of Technology, 2008(12):154-159.
[4] Zhang Xiangxu [1] , Zheng Shixiong [1] , Tang Wei [2] , et al. Numerical Simulation of Wind Parameters in Long Span Arch Bridges in Mountain Canyons[J]. Journal of Railway Science and Engineering, 2018.
[5] Uchida T , Ohya Y . Large-eddy simulation of turbulent airflow over complex terrain[J]. Journal of Wind Engineering and Industrial Aerodynamics, 2003, 91(1-2):219-229.
[6] SMAGORINSKY J. General circulation experiments with the primitive equations[J]. Monthly Weather Review, 1963, 91(3)： 99-164.
[7] LAUNDER B. Low Reynolds number turbulence near walls[R]. Manchester： University of Manchester, Report TFD/86/4, 1986.
[8] Durbin P A . Separated flow computations with the k-epsilon-v-squared model[J]. 2012.
[9] Boussinesq, J. (19877), "Essai sur la théorie des eaux courantes", Mémoires présentés par divers savants à l'Académie des Sciences23(1): 1-680
[10] Wang M , Chen Q . On a Hybrid RANS/LES Approach for Indoor Airflow Modeling (RP-1271)[J]. HVAC&R Research, 2010, 16(6):731-747.
[11] LI Yongle, HU Peng, CAI Xian-xi, et al. Numerical simulation of wind characteristics in the vicinity of high-steep mountain bridge site[J]. Journal of Aerodynamics, 2011, 29(6): 770-776.

Acknowledgments
The authors would like to acknowledge with great gratitude for the supports of the national science foundation of China (grant no: 51778551), the science and technology plan project of Xiamen (grant No: 3502Z20161016).

IOP Conf. Series: Earth and Environmental Science **356** (2019) 012013 doi:10.1088/1755-1315/356/1/012013

Directional information coupling dynamics in complex climate system

LH Yuan[1] and YJ Zhong[2]

1 Southern University of Science and Technology, Shenzhen, China
E-mail: 11510830@mail.sustech.edu.cn
2 Chongqing Jiaotong University, Chongqing, China
E-mail: matthew.ambrosius@outlook.com

Abstract Climate change has been significantly disturbing the dynamics of different earth system components, such as land surface and ocean, as well as the interactive relationship between different components. Here we aim to investigate the sea surface temperature dynamics and its remote connection to the precipitation patterns. Previous efforts on the remote ocean-land remote coupling are mostly relied on linear based statistical inference framework, disregarding the nonlinearity of the earth system dynamics. Here we apply a new inference framework that fully adapts to nonlinear system to quantify the coupling strength between Atlantic Oceanic temperature signals (AMO index) and US precipitation patterns. We found that the linear based coupling patterns are significantly different from the nonlinear based coupling patterns, which provides important insights into the system nonlinearity. We also conduct uncertainty analysis to quantify the estimated coupling strength uncertainty and discuss the robustness of the climate coupling between AMO and US precipitation.

1.Introduction

Climate warming has been ongoing since the pre-industrial era, due to the vast amount of greenhouse gas emissions and the positive radiative forcing of the potent greenhouse gases, such as CO_2 and CH_4. However, the warming over the ocean and land surfaces is neither evenly distributed, nor independently developed. Intrinsic relationship between land surface and sea surface temperature warming provides important insights into how the coupled earth system climate evolve dynamically, and how remote oceanic-land temperature signals relate to one another. Here, we particularly focus on the relationship between the Atlantic Sea Surface Temperature (SST) and United States land surface temperature. For the oceanic SST, the Atlantic Multidecadal Oscillation (AMO) index is considered, which has been known as the North Atlantic sea surface temperatures (SSTs) oscillating between high and low temperatures with a period of 60–80 years over a multidecadal timescales [1][2]. The AMO index of North Atlantic SST was calculated by averaging the low-pass filtered annual mean SST Anomalies over the region 0–60 N, 75 W–7.5 W [3]. A number of studies have shown that AMO have strong impacts to global and regional climate changes. For example, the AMO is associated with droughts and rainfall frequency over the conterminous United States. It may have a crucial influence on the teleconnection between the El Niño Southern Oscillation (ENSO) and winter precipitation [4]. AMO has been proposed to play an important role in the decades of changes in summer rainfall in India and the Sahel in the 20th century and is critically relevant to Atlantic hurricane activity [1]. The opposite phase of AMO anomaly from 1950 to 2015 corresponds to the significant difference in activity of atmospheric action and regional zonal circulation and heat and moisture transfer from the Atlantic to the continent [5]. North Atlantic Ocean have been suggested to be a key driver of European

Content from this work may be used under the terms of the Creative Commons Attribution 3.0 licence. Any further distribution of this work must maintain attribution to the author(s) and the title of the work, journal citation and DOI.
Published under licence by IOP Publishing Ltd

climate shift. AMO is remarkably consistent with anomalies in surface air temperature (SAT), sea level pressure (SLP) and precipitation in European [6].

Robustly investigating the relationship between AMO and land surface temperature is challenging due to the complexity and nonlinearity of earth system dynamics. Previous efforts have applied many different statistical inference frameworks to tackle this question. For example, Yu et al. applied empirical-orthogonal-function (EOF) analyses to the seasonal warm anomalies for getting the principal components (PCs). Based on the Linear correlation analysis between the PCs and the AMO index, they find that AMO can make a significant contribution to seasonal warm anomalies across the contiguous U.S. in the three most recent decades through large-scale atmospheric circulation [7]. However, correlation coefficients are not complete to describe correlation between variables since it just measures the linear dependence. In this work, we aim to overcome the challenge of nonlinearity in studying the complex relationship between AMO and land surface temperature dynamics. We propose to use the information entropy, which to quantify the amount of uncertainty involved in the value of a random variable or the outcome of a random process [8]. Particularly, the mutual information measures the information shared between two processes, which account for nonlinear dependence. we suggest that it is more suitable to apply the mutual information to evaluate the non-linear coupling between the AMO index and the land temperature or precipitation, compared with linear correlation-based analysis.

2. Methodology

2.1.Data description

We obtain Sea Surface Temperature (SST) for the region that define AMO Index Data from the Climate Analysis Section, NCAR, Boulder, USA, Trenberth and Shea (2006) [9], accessed January 30, 2019. We use the detrended AMO that averaged over the region 0-80N from 1958 to 2013 [6]. For precipitation data, we use the monthly Global Precipitation Climatology Center (GPCC) product ranging from 1958 to 2013 (56 years) [10]. Below, Figure 1 shows the detrended monthly averaged AMO index, which covers a relative dominated cold phase from about 1959 to 1995. Figure 2 shows the United States precipitation annual mean and standard deviation. Precipitation rates are high over the western coastal area and the southeast part of US.

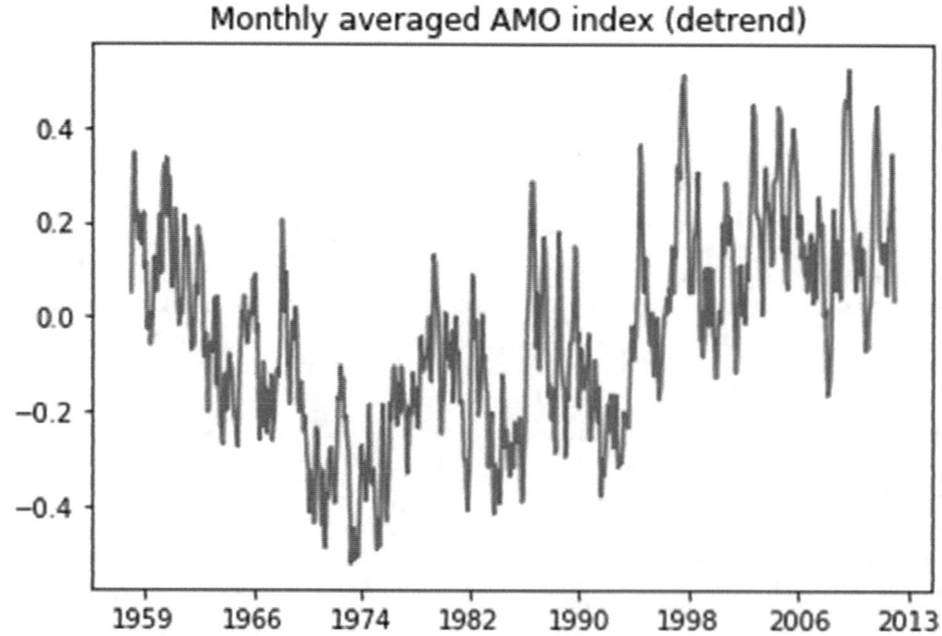

Figure 1 The detrended monthly averaged AMO index for the period of 1958-2013

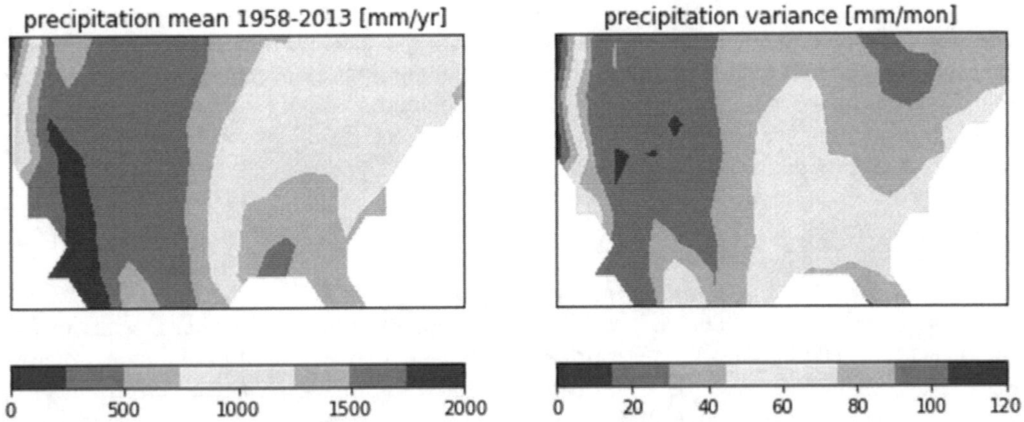

Figure 2 Distribution map of annual precipitation mean 1958-2013 in United States(left). Unit: mm/year Distribution map of precipitation variance in the United States. Unit: mm/mon

2.2. Linear statistical model
We first use Pearson correlation coefficient to investigate the coupling relationship between AMO and US precipitation at monthly time scale.

$$\rho_{X,Y} = \frac{\operatorname{cov}(X,Y)}{\sigma_X \sigma_Y} \quad (1)$$

where $\operatorname{cov}(X,Y) = E[(X-\mu_X)(Y-\mu_Y)]$ is the covariance, *E* is the expectation. μ_X is the mean of *X*, μ_Y is the mean of *Y*. $\boldsymbol{\sigma}_X$ is the standard deviation of *X*, $\boldsymbol{\sigma}_Y$ is the standard deviation of *Y*. The Pearson correlation has a value of between +1 and -1, where 1 is positive correlation, 0 is no linear correlation, while -1 is negative correlation.

2.3.Non-linear statistical model
We also use the mutual information derived from Shannon information entropy theory [9] to quantify the relationship between AMO and US regional precipitation dynamics.

$$H(X) = -\sum_x p(x)\log_2 p(x) \quad (2)$$

$$H(X|Y) = -\sum_{x,y} p(x,y)\log_2 p(x|y). \quad (3)$$

$$I(X;Y) = \sum_{x,y} p(x,y)\log_2 \frac{p(x,y)}{p(x)p(y)}. \quad (4)$$

where Eqn. 2 define the Shannon entropy *H,* in the units of bits as it use a logarithm of base. *P(x)* is the probability of the source process *X* to occur. The conditional entropy of two processes *X* and *Y* is given by Eqn. 3. Where *p(x,y)* is the joint distribution, *p(x|y)* is the conditional distribution of *X* given *Y*. The mutual information of *X* relative to *Y* is given by Eqn. 4. The mutual information measures mutual dependence between two variables which contain linear and nonlinear dependence [9]. When the special case that absolute values of Pearson correlation coefficient equal to 1, which means X and Y have a perfect linear correlation. In this case, the mutual information of *X* relative to *Y* are given as: I(X;Y)=H(X)=H(Y).

IOP Conf. Series: Earth and Environmental Science **356** (2019) 012013 doi:10.1088/1755-1315/356/1/012013

2.4.Analysis
In this study, we utilize two different types of statistical tools: linear and Non-linear inferences. Given the AMO index as the X variable and the us precipitation data is the Y variable. First, we apply the Pearson correlation coefficient equation (Eqn. 1) to analyse the linear correlation between AMO index data and American land precipitation data. Second,we evluate the non-linear mutual information by measureing mutual information of AMO index relative to US percipatation data is given by Eqn. 4. Accurate estimation of the marginal possibility distribution function (pdf) of X and Y is crucial to eastimate mutual information. We have used Ordinary ranking (ORD) and Gaussian kernel estimation (GKE) to estimate possibility distribution function.

3. Results and Discussion

3.1. Probability density estimate for mutual information calculation
Probability density distribution (PDF) of AMO and precipitation time series are estimated by Ordinary ranking (ORD) and Gaussian kernel estimation (GKE) with different bin numbers. As shown in the Figure 3, PDF obtained by ORD is significantly affected (high sensitivity and uncertainty) by the change in bins value. By contrast, the PDF obtained with GKE is relatively stable as bins' values equal to 9,10,11,12. Therefore, we conclude that GKE is a more stable and reliable PDF estimator for our analysis.

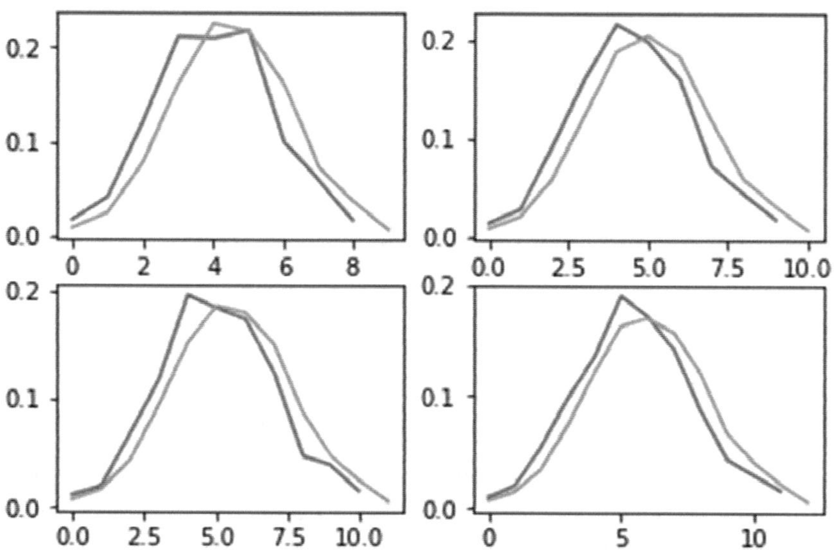

Figure 3 The possibility distribution function of normal distributed random samples estimated by Gaussian kernel estimation (GKE) and Ordinary ranking (ORD). bin=9,10,11,12 for top left, top right, bottom left, bottom right panels.

3.2. Linear versus nonlinear relationship inference
Pearson Correlation coefficient is a measure of linear and non-directional relationship between two variables. But Mutual information is a measure of directional non-linear causality. Here we quantify the spatial patterns of Pearson correlation coefficient (Figure 4 left panel) and mutual information (Figure 4 right panel) between the AMO and precipitation dynamics over the United States. Figure 4 shows that the two spatial pattern are not consistent.In general, the linear dependency between AMO and US precipitation is weak ($r^2<0.2$), while the non-linear relationship between AMO and US precipitation is relative stable and statistically significant across a large part of US (colorred regions all pass statistical significance test). Comparing the spatial distribution, the western United States, especially the southwestern region, has a strong non-linear relationship, while very little linear

dependency. This indicates a very strong nonlinear coupling relationship between the remote AMO and precipitation over this region, which may not have been identified and considered in previous studies, due to the statistical inference method limitation. In northern and central United States, both the linear and nonlinear coupling relationships are prominent, which is consistent with what previous studies suggested that AMO had a significant role in determining the circulation and participation in this region. [11] One possible reason is that our data covers a significant long AMO cold phase.

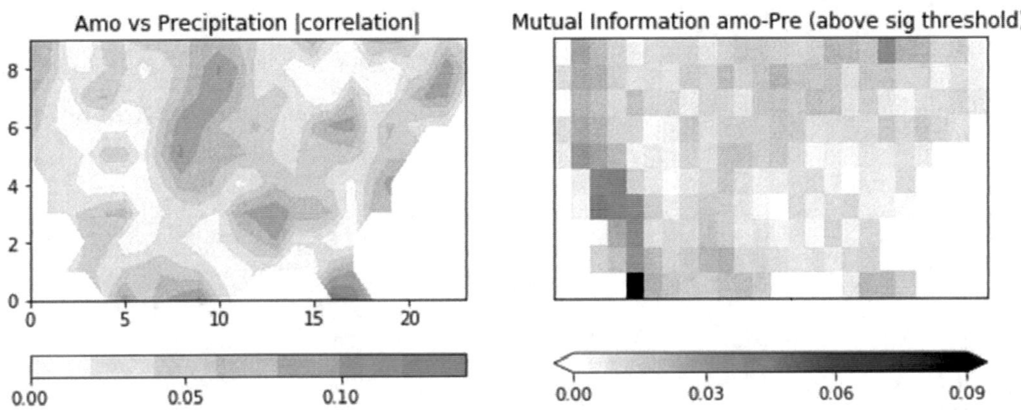

Figure 4 Spatial pattern of the Pearson correlation coefficient and Mutual information (PDF=GKE, bin=14) between AMO and precipitation in the United States.

3.3. Climate change effects on ocean-land coupling

Here we divide the AMO time series data into two equal segments, each has 28-year: 1958-1985 and 1986 -2013. Since the 20th century, the average temperature of the earth's climate system has risen rapidly. It will increase the amount of evaporation and cause more precipitation. We attempted to quantify the effect of temperature rise on the nonlinear coupling between the Atlantic and continental precipitation in the United States by comparing the first and second 28-year AMO and precipitation mutual information fluctuations. Figure 5 shows that the mutual information between the AMO and precipitation have not changed significantly. Which means that global warming does not have a detectable impact on this coupling strength between the two processes of interest.

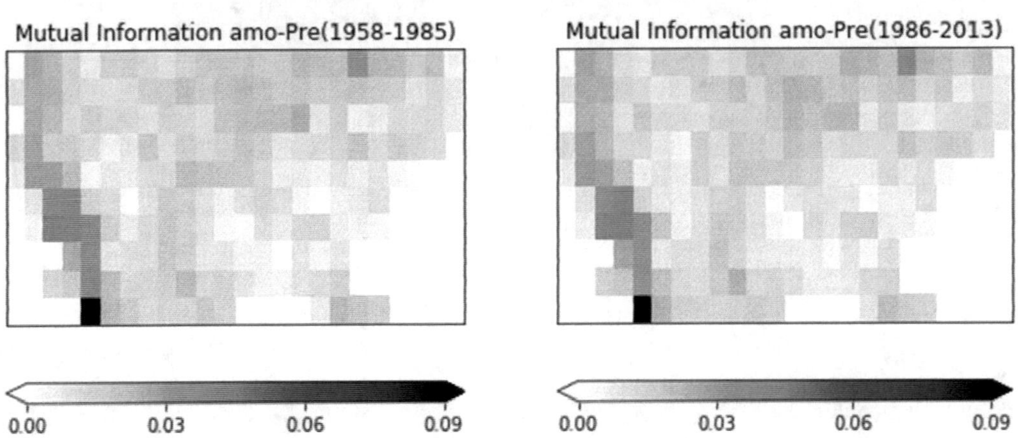

Figure 5 Spatial pattern of the Mutual information between amo and precipitation in the United States from 1958-1985(left) and 1986-2013(right). (PDF=GKE, bin=14)

3.4. Uncertainty analysis

The mutual information between AMO and precipitation will be affected by some uncertain factors, such as our choice of Probability Density Function (PDF) and Bin value. We conduct uncertainty analysis to show in Figure 6, that the mutual information between AMO and precipitation are subject to uncertainties. Figure 6 top panels use Ordinary Ranking (ORD) as the PDF estimator, and the Bin numbers are set to 10, 12, and 14, respectively. As the bin number increases, the identified nonlinear coupling relationship become stronger. The bottom panel of Figure 6 selects GKE as the pdf. As the Bin value increase, the change in the mutual information is limited, reflecting the stability of identified spatial distribution of mutual information. Consistent with our analysis in Figure 3, GKE is more stable as a PDF, which helps us to obtain a more accurate assessment of the nonlinear coupling between AMO and precipitation. By comparing the spatial maps of the top and bottom panels, it can be seen that the mutual information are different and he mutual information value calculated by ORD is higher.

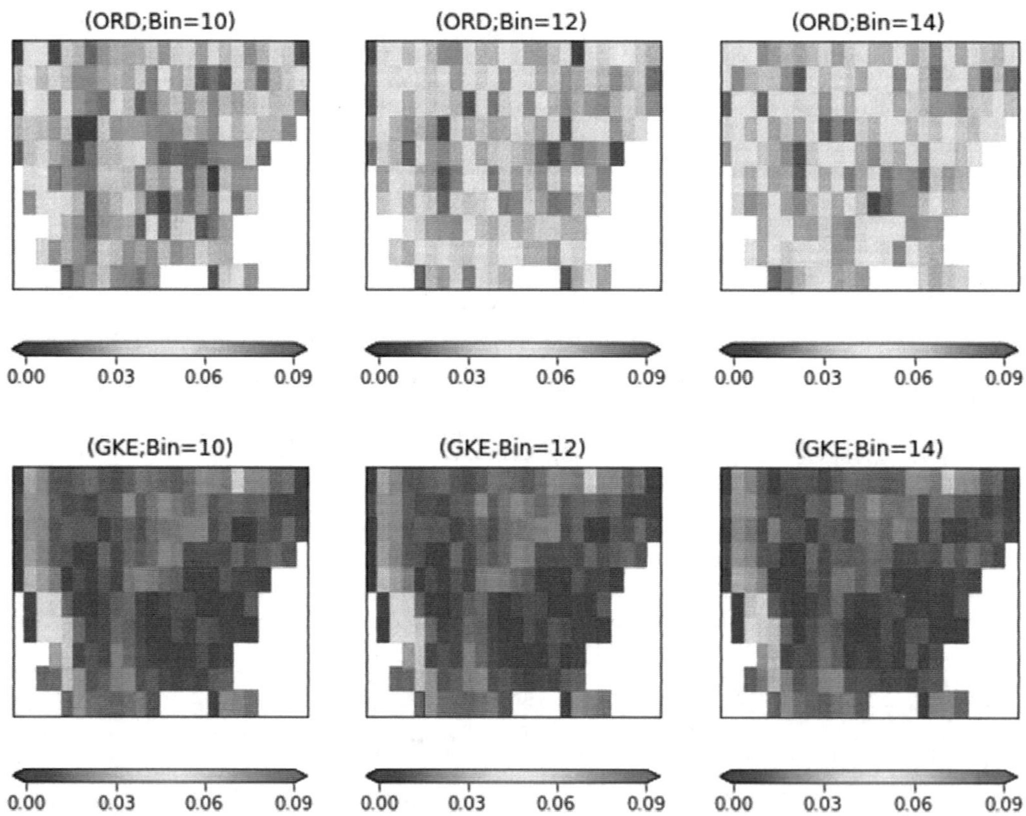

Figure 6 Spatial pattern of the Mutual information between AMO and precipitation in the United States.PDF=GKE or ORD; Bin=10 or 12 or 14.

4. Conclusions

In this study, we quantify the coupling relationship between Atlantic Ocean surface temperature (AMO) and the US regional precipitation, through linear and non-linear inferences. We find that the linear relationship between AMO and US precipitation are generally weak, while the nonlinear relationships are statistically significant and particularly strong over the southwestern US. Such a strong nonlinear coupling relationship in the southwestern US is complex and likely associated with other important climatic mode ,i.e., ENSO, PDO and SOI. Previous study suggested that strong ENSO teleconnections with participation in the western US appears associated with the negative PDO phase and out of SOI phase which characterize the interaction of tropical air and sea circulation. [12] To understand the nonlinear coupling relationship between the AMO index and the participation furtherly, there still need more studies to figure out physical processes and the associations between various climate modes. In northern and central United States, previous studies suggested that positive precipitation anomalies have been observed in this region during the negative phase of the AMO. [11] Such result also indicated in our linear analysis probably since our time series data is dominated by a cold phase in this region. However, the nonlinear coupling are also statistically significant over northern and central US, which has not been identified in previous studies.Moreover, climate warming during the historical period is not strong enough to significantly alter such non-linear coupling relationship. Finally, our uncertainty analysis reveal that the non-linear coupling patterns are robust using GKE probability density function estimator, while less robust using ORD probability density function estimator, the latter is sensitive to the bin numbers and does not clearly show spatial distinctions in the coupling strength over the US regions.

5. References

[1] Zhang, R., & Dilworth, T. L. (2006). Impact of Atlantic multidecadal oscillations on India/Sahel rainfall and Atlantic hurricanes. *Geophysical Research Letters, 33*(17), 123-154.

[2] Kurtz BE (2015) The Effect of Natural Multidecadal Ocean Temperature Oscillations on Contiguous U.S. Regional Temperatures. PLoS ONE 10(6): e0131349. doi:10.1371/journal.pone.0131349

[3] Sutton, R. T. , & Hudson, D. L. R. . (2005). Atlantic ocean forcing of north American and European summer climate. *Science, 309*(5731), 115-118.

[4] McCabe, G. J., Palecki, M. A., & Betancourt, J. L. (2004). Pacific and Atlantic Ocean influences on multidecadal drought frequency in the United States. *Proceedings of the National Academy of Sciences of the United States of America, 101*(12), 4136-41.

[5] Semenov, V. A. , & Cherenkova, E. A. . (2018). Evaluation of the atlantic multidecadal oscillation impact on large-scale atmospheric circulation in the atlantic region in summer. *Doklady Earth Sciences,478*(2), 263-267.

[6] Sutton, R. T. , & Dong, B. . (2012). Atlantic ocean influence on a shift in european climate in the 1990s. *Nature Geoscience*.

[7] Yu, L. , Zhong, S. , Heilman, W. E. , & Bian, X. . (2018). Trends in seasonal warm anomalies across the contiguous united states: contributions from natural climate variability. *Scientific Reports, 8*(1), 3435.

[8] Schneider et al (2017): Evaluating the Hydrological Cycle over Land Using the Newly-Corrected Precipitation Climatology from the Global Precipitation Climatology Centre (GPCC). Atmosphere 8(3), 52; doi:10.3390/atmos8030052

[9] Leong, K. Y., Saidur, R., Khairulmaini, M., Michael, Z., & Kamyar, A. (2012). Heat transfer and entropy analysis of three different types of heat exchangers operated with nanofluids. International Communications in Heat and Mass Transfer, 39(6), 838-843.

[10] AMO Index Data provided by the Climate Analysis Section, NCAR, Boulder, USA, Trenberth and Shea (2006). Accessed January 30, 2019.

[11] Hu, Q. , Feng, S. , & Oglesby, R. J. . (2011). Variations in north american summer precipitation driven by the atlantic multidecadal oscillation. *Journal of Climate, 24*(21), 5555-5570.

[12] Mccabe, G. J. , & Dettinger, M. D. . (1999). Decadal variations in the strength of enso teleconnections with precipitation in the western united states. *International Journal of Climatology, 19*(13), 1399-1410.

IOP Conf. Series: Earth and Environmental Science **356** (2019) 012014 doi:10.1088/1755-1315/356/1/012014

Study on Preparation Conditions of Electrolyzed Reduced Alkaline Water

liuxin Jing [1], mengliang Chen [2] and huaiyu Cai*

1 Chinese Academy of Environmental Planning, Beijing, China
2 China municipal engineering northeast design research Institute limited, Changchun, China
E-mail: 349646725@qq.com

Abstract. In this study, the preparation conditions (electrolytic voltage, electrode interval and electrolysis time) of electrolyzed reduced alkaline water were optimized. Results show that electrolytic time, electrolytic voltage and electrode interval could influence the water quality of the electrolyzed reduced alkaline water. More electrolysis time, higher of electrolytic voltage, and smaller electrode interval resulted in higher water pH value. More electrolysis time, higher electrolytic voltage, and larger electrode interval resulted in lower water ORP value. The optimum condition for preparation of weak alkaline primary water is 10v electrolytic voltage, 15 mm interval of electrodes and 2 min electrolysis time.

1. Introduction

Safety of drinking water is the basic requirement for a beautiful life. The development of water treatment technology has laid a solid foundation for the drinking water treatment [1]. The drinking water quality played a very positive role in health of humans. With the increasing of complex pollutions were discharged in to environment, conventional drinking water treatment process is difficult to ensure the quality of drinking water, therefore, the problem of drinking water pollution is becoming more and more serious [2]. In order to achieve the goal of safer and better drinking water, some new advanced treatment technologies, such as electrolysis technology, electromagnetic treatment, mineralization and other technologies, have been developed rapidly [3]. The drinking water produced by electrolysis technology is welcomed by consumers was reported to improve human physiological function [4], like eliminate internal redundant free radicals, prevent oxidative damaging [5] and reduce blood cell adhesion and aggregation and remit early coagulation [6].In literatures, this new type of electrolyte drinking water was named as alkaline ion water or weak alkaline ion water [7] [8] [9]. In our study, weakly electrolyzed reduced alkaline water was used according to its characteristics like pH, redox potential and preparation method. The characteristics of this weakly electrolyzed reduced alkaline water were firstly reported by several Japanese scientists in the 1980s. They firstly developed the produce process to quantity produce this weakly electrolyzed reduced alkaline water [10]. Due to water quality and storage problem [11], however, that method did not world widely spread. At present, many devices were invented to produce this water and the household electrolytic water machine is one of the most commonly used devices for the preparation of weakly electrolyzed reduced alkaline water. The main structure of this machine generally includes power supply, electrolyzer, electrode and selective penetration membrane. The produced water quality and quantity is mainly depended on the structure and operation of the preparation device. This study simulated the water production process of the household electrolytic water machine and explored the best preparation conditions.

Content from this work may be used under the terms of the Creative Commons Attribution 3.0 licence. Any further distribution of this work must maintain attribution to the author(s) and the title of the work, journal citation and DOI.
Published under licence by IOP Publishing Ltd

2. Material and methods

2.1. Experiment setup and operation procedure
In this experiment, a self-designed and assembled weakly electrolyzed reduced alkaline water generating device was adopted which can effectively control the operating condition. As Figure 1, the device mainly includes three parts: power, pretreatment device (water purifier) and electrolyser. During the experiment, municipal tap water was firstly poured into water purifier and then injected into the electrolyzer (Figure 2).

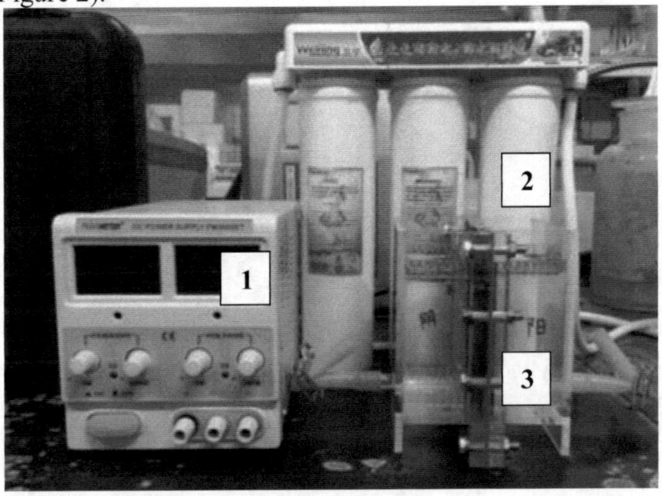

1. Power, 2. water purifier, 3. electrolyzer

Figure 1. Physical diagram of self-made low alkaline electrolytic primary water generating device.

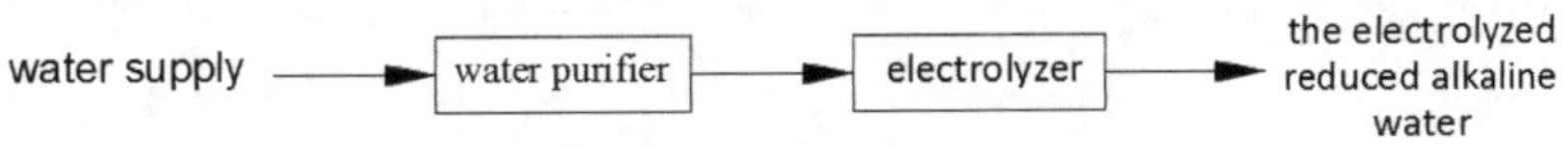

Figure 2. Flow chart of preparation

2.2. Single factor method
Single factor method was used to optimum preparation conditions to produce weakly electrolyzed reduced alkaline water. The effect of electrolytic voltage, electrolytic time and electrode interval on water quality was respectively analyzed by single factor method, as follows:
Under electrode interval of 10 mm and electrolytic time of 1 min, the voltage was adjusted from 5v, 10v, 15v, 20v, 25v to 30v, respectively. After that, pH and ORP value of the water were measured. The effect of electrode spacing (10 mm, 15 mm, 20 mm, 25 mm and 30 mm) on water quality was studied under voltage 10 v and the electrolytic time 1min. In the experiment of the effect of electrolysis time (1 min, 2 min, 3 min, 4 min and 5 min) on the water quality with the electrode interval 10 mm and the electrolytic voltage 10 V.

2.3. Orthogonal experiment method
Optimum preparation conditions of electrolyte water were conducted by orthogonal method according as following describe:
Standard orthogonal table L9 (34) was selected, and the orthogonal experiment design assistant was used to carry out the experiment design. The specific factors and horizontal distribution are shown in Table 1, and the experimental scheme is arranged as shown in Table 2.

Table 1. The orthogonal experiment in three factors and three levels.

Level	Factor		
	Electrolytic voltage (V)	Electrode interval (mm)	Electrolytic time (min)
1	Ten	Ten	1
2	Fifteen	Fifteen	2
3	Twenty	Twenty	3

Table 2. The orthogonal experimental design method in three factors and three levels.

Serial number	Factor			
	A Electrolytic voltage (V)	Empty column	B Electrode interval (mm)	C Electrolytic time (min)
1	1 (10)	1	1 (10)	1 (1)
2	1	2	2 (15)	2 (2)
3	1	3	3 (20)	3 (3)
4	2 (15)	1	2	3
5	2	2	3	1
6	2	3	1	2
7	3 (20)	1	3	2
8	3	2	1	3
9	3	3	2	1

2.4. Response surface method

In this experiment, response surface method was used to verify the regression relation between key influencing factors (electrolytic voltage, electrode interval and electrolytic time) and electrolysis conditions.

In this paper, using MATLAB software, and using the data obtained from single factor experiment and orthogonal experiment, we set up a suitable mathematical model, and analyze the experiment index (pH, ORP) and electrolysis conditions. According to their regression relation, draw a corresponding relation graph.

2.5. Analytical methods

pH and ORP were chosen to evaluate water quality of electrolyzed reduced alkaline water which were measured by the HI98194 portable multi-parameter water quality analyzer (HANA, Italy). All of the experiments and analysis were conducted in triplicate and the results were expressed in average. The experiment data is processed by using the Origin 9.0 (OriginLab, US).

3. Results and discussions

3.1. Influence of electrolytic voltage

Table 3. Different electrolytic voltage and corresponding discharge index.

V (V)	5	Ten	Fifteen	Twenty	Twenty five	Thirty
pH	7.7	8.37	8.9	9.17	9.33	9.50
ORP (mV)	66	38	34	11	9	-16

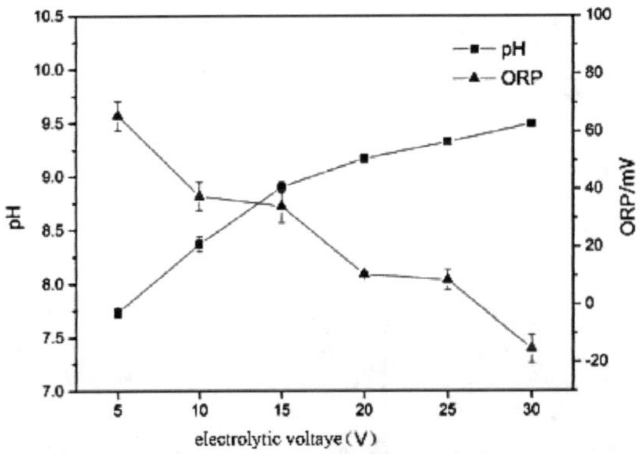

Figure 3. Influence of different electrolytic voltage on water quality

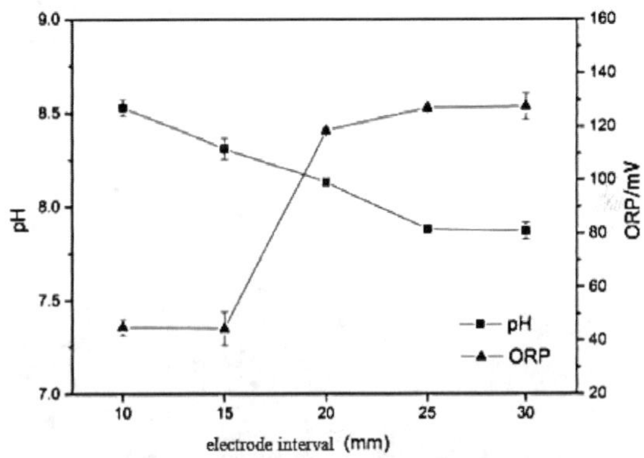

Figure 4. Influence of different electrode spacing on water quality

Table 3 and Figure 3 shows the pH and ORP values of the electrolyzed reduced alkaline water with voltage increased from 5V to 30V. The electrolytic voltage has great influence on the pH and ORP values of electrolyzed reduced alkaline water. pH value increases continuously with the increase of electrolytic voltage especially when the voltage lower than 15V. The pH value increased slowly when voltage higher than 15V, and kept around 9.50 under 30V. The ORP value exhibited a conversely variation trend compared with pH which was decreased from 66 mV to -16 mV. According to the pH and ORP variation, the suitable voltage value was between 10V and 20V, and this voltage range is agreement with the limit voltage range (between 3V and 2V).

3.2. Influence of electrode interval

Table 4. Different electrode spacing and corresponding discharge index.

H (mm)	Ten	Fifteen	Twenty	Twenty five	Thirty
pH	8.53	8.31	8.13	7.88	7.87
ORP (mV)	45	44.5	118.5	127	127.5

Under the condition that the electrolytic voltage was 10V and the electrolytic time was 1 min, the influence of electrode interval was shown in Table 4 and Figure 4. According to the above experiment results, the electrode interval also influenced pH and ORP values of electrolyzed reduced alkaline water. With the increase of the pole interval, pH slightly decreased from 8.53 to 7.87 and the ORP significantly increased from 45 mV to 127.5 mV. It can be concluded that decrease the distance of electrodes was benefit for the quality of produced electrolyzed reduced alkaline water. However, the small electrode distance means less water between two electrodes, which would result in a lower ion exchange effect. After comprehensive consider the water quality and water quantity, the pole interval between 10 mm and 20 mm was thought as the most suitable.

3.3. Effect of electrolysis time

Table 5. Different electrolysis time and corresponding discharge index.

T (min)	0	1	2	3	4	5
pH	7.56	8.42	8.99	9.32	9.34	9.38
ORP (mV)	176	41	8	-35	-75	-112

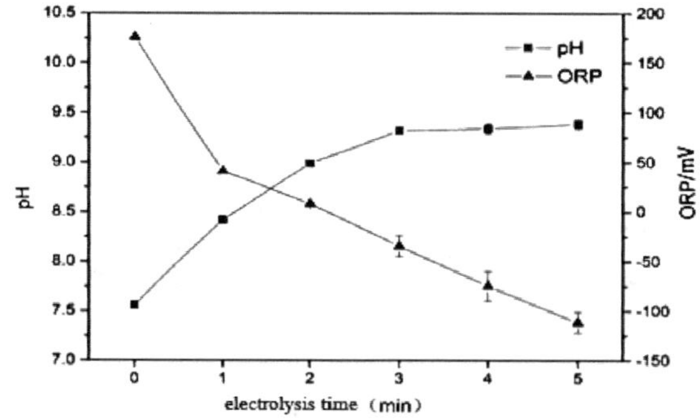

Figure 5. Effect of different electrolysis time on water quality

Table 5 and Figure 5 listed the influence of electrolytic time on the electrolyzed reduced alkaline water quality. From Figure 5, electrolysis time has great influence on pH and ORP values of electrolyzed reduced alkaline water. The pH significantly increased from 7.56 to 9.32 in the first 3 min, and after that, pH value kept stable around 9.3 with electrolysis time increased to 5 min indicating that raw water was completely ionized in 3 min. The ORP value kept decrease during 5 min. In 1 min, the pH value can reached low alkaline condition, while the negative ORP value need more time (3 min). It can be seen that to prepare the electrolyzed reduced alkaline water the electrolysis time should more than 1 min. Comprehensive electrode life and energy consumption, the optimum electrolysis time is 1-3min for the electrolysis device.

3.4. Best preparation conditions

3.4.1. Analysis of orthogonal experiment. According to the results of electrolytic voltage, electrode interval and electrolytic time on the quality of electrolyzed reduced alkaline water, the suitable range of electrolytic conditions was listed in Table 6, and the visual analysis results are shown in Table 7.

Table 6. Three factors three level orthogonal experiment results.

Serial number	Factor	Experiment indicators	Experimental phenomena

	A Electrolytic voltage (V)	Empty column	B Electrode interval (mm)	C Electrolysis time (min)	pH	ORP	Bubble	Precipitate
1	Ten	1	Ten	1	8.53	-8.6	Nothing	Nothing
2	Ten	2	Fifteen	2	9.07	-45.65	Trace	Trace
3	Ten	3	Twenty	3	9.01	-34	Small amount	Nothing
4	Fifteen	1	Fifteen	3	9.31	-65	Large	Small amount
5	Fifteen	2	Twenty	1	8.66	-4.2	Small amount	Nothing
6	Fifteen	3	Ten	2	9.35	-70.75	Large	Small amount
7	Twenty	1	Twenty	2	9.24	-37.75	Small amount	Small amount
8	Twenty	2	Ten	3	9.86	-109.8	Large	Large
9	Twenty	3	Fifteen	1	8.94	-28.35	Large	Nothing

All of the nine conditions in Table 6 could produce electrolyzed reduced alkaline water with low alkalinity pH value (8.5 and 9.5) and negative ORP value indicated that the range of electrolytic conditions selected in Table 6 is suitable.

Table 7. Three factors-level visual analysis results.

Factor	A Electrolytic voltage(V)	Empty column	B Electrode interval (mm)	C Electrolysis time(min)
K1 (pH)	8.87	9.03	9.25	8.71
K2 (pH)	9.11	9.20	9.11	9.22
K3 (pH)	9.35	9.10	8.97	9.39
R (pH)	0.48	0.17	0.28	0.68
K1 (ORP)	-29.42	-37.12	-63.05	-13.72
K2 (ORP)	-46.65	-53.22	-46.33	-51.38
K3 (ORP)	-58.63	-44.37	-25.32	-69.60
R (ORP)	29.22	16.10	37.73	55.88

According to the range analysis of pH value and ORP value in Table 7, for the pH value, the influence of electrolytic time was largest following with electrolytic voltage and electrode interval. For water ORP value, the electrolytic time was the most important influencing factor following with electrode interval and electrolytic voltage. The empty column has very little effect compared with other factors; therefore the interaction can be ignored in this experiment.

The principle of optimum preparation conditions selection, on one hand, the water pH value is the higher the better (should less than 9.5) while the ORP value is the lower the better. On the other hand, the energy consumption should be consideration in practice. And from the user experience, electrolysis time should be as short as possible. Therefore, A1B2C2 was thought as the optimum scheme in this experiment, that is to say, the electrolytic voltage is 10v, the electrode spacing is 15 mm and the electrolytic time is 2 min.

3.4.2. Analysis of response surface. The pH and the quadratic regression equations of the electrolyzed reduced alkaline water were established by MATLAB software:

(a) Analysis of the responsiveness of pH and the electrolytic conditions:

The pH and the quadratic regression equations of the electrolyzed reduced alkaline water were established by MATLAB software:

$$D=7.5101+0.0954A+0.0319B+0.2017C-0.0002A^2+0.0008B^2-0.0142C^2-0.0045AB+0.0113AC-0.0051BC \quad (1)$$

in which D is pH, A is electrolytic voltage (V), B is electrode spacing (mm), and C is electrolytic time (min).

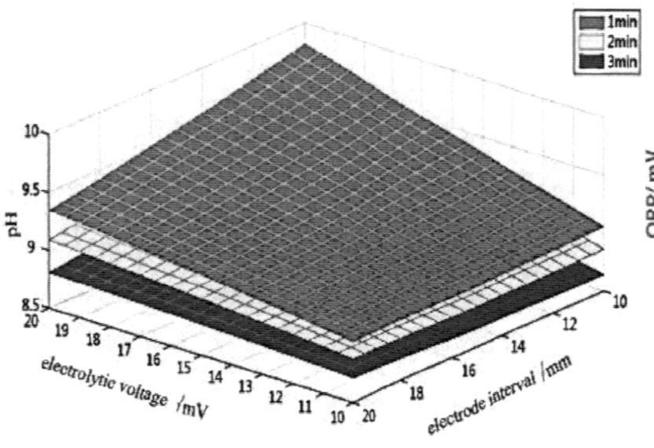

Figure 6. The relationship between pH and electrolytic condition by response surface analysis

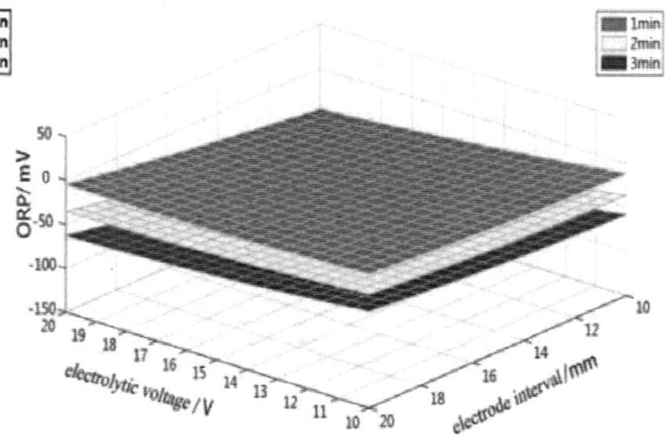

Figure 7. The relationship between ORP and electrolytic condition by response surface analysis

At the same time, three statistics of the model are obtained(Figure 6): the fitting degree $R^2=0.9391$, the statistical observations value F=15.4079, and the examine P=0.00018. It is known that when the R^2 is greater than 0.9 and closer to 1, and the F value is greater, the model fitting degree is higher and the correlation is better. When p value less than 0.05 illustrate that the model is effective. Therefore, the above model is effective and has high similarity and correlation.

(b) Analysis of the responsiveness of ORP and the electrolytic conditions:

The ORP and the quadratic regression equations of weak alkaline electrolytic primary water using MATLAB software are as follows:

$$D=102.8266-9.0280A-4.4568B-24.2071C+0.1154A^2+0.0952B^2+1.7188C^2+0.2811AB-0.7671AC+0.1796BC \quad (2)$$

where D is ORP value, A is electrolytic voltage (V), B is electrode spacing (mm), and C is electrolytic time (min).

At the same time, three statistics obtained from the above model (Figure 7): the fitting degree R2=0.9568, the statistical observations value F= 22.1394 and the examine P=0.00004. It is known that when the R2 is greater than 0.9 and closer to 1, and the F value is greater, the model fitting degree is higher and the correlation is better. When p value less than 0.05 illustrate that the model is effective. Therefore, the above model is effective and has high similarity and correlation.

4. Conclusion

(a) Electrolytic time, electrolytic voltage and electrode interval could influence the water quality of the electrolyzed reduced alkaline water. More electrolysis time, higher of electrolytic voltage, and smaller electrode interval resulted in higher water pH value. More electrolysis time, higher electrolytic voltage, and larger electrode interval resulted in lower water ORP value.

(b) The optimum condition for preparation of weak alkaline primary water is 10v electrolytic voltage, 15 mm interval of electrodes and 2 min electrolysis time.

(c) High fitting degree and correlation regression relations were established between water quality (pH and ORP values) and operation conditions (electrolytic voltage (A), electrode interval (B) and electrolytic time (C)):

pH: $D=7.5101+0.0954A+0.0319B+0.2017C-0.0002A^2+0.0008B^2-0.0142C^2-0.0045AB+0.0113AC-0.0051BC$

ORP: $D=102.8266-9.0280A-4.4568B-24.2071C+0.1154A^2+0.0952B^2+1.7188C^2+0.2811AB-0.7671AC+0.1796BC$

5. References

[1] Riming L, HuangJian, Tingjin Y, et al 2012 Actuality and Progresson Treatment Technology of Drinking Water. *J. Water and Wastewater Engineering* **S1** 213-218

[2] Jinlong Z. 2007 Evaluation and technology integration of drinking water treatment technology *D. Harbin Institute of Technology*

[3] Xixuan Z 2009. Analysis on current situation and safeguard countermeasures of drinking water safety in China *J. Environment and Sustainable Development* **10** 221-221

[4] Qing S 2012.Molecular acid-base chemistry. Shanghai :*Shanghai science and Technology Literature Press*

[5] Shirahata S； Kabayama S； Nakano M et.al 1997. Electrolyzed-reduced water scavenges active oxygen species and protects DNA from oxidative damage *J. Biochemical and Biophysical Research Communications* **01** 269-274

[6] Rubik B 2011. Studies and observations on the health effects of drinking electrolyzed-reduced alkaline water *C. Water and Society* 317-327

[7] Shuqin L and Haiping L 2006. Preparation principle, characteristics and application of alkaline ionic water *J. Journal of Changzhi University* **02** 44-45.

[8] Yonghong M 2002.Drinking water and fatigue recovery after exercise- Test and analysis of the blood HL of athletes after training with weak alkaline electrolyzed reduction water *J. Journal of Beijing Sport University* **04** 480-481.

[9] Yongyuan S 2002.Characteristics and functions of ionic water *J. Science and Technology of Food industry* **01** 74-76.

[10] Henry M, Chambron J. Physico-Chemical 2013. Biological and Therapeutic Characteristics of Electrolyzed Reduced Alkaline Water (ERAW) *J. Water* **04** 2094-2115.

[11] Zhongyi L, Qibin L and Miaoling Z 2002. Ionic water process and quality improvement method *J. The Beverage Industry* **05** 27-30.

An empirical research on the relationship between water quality,climate change and economic development in Jiulong River Watershed

Guangsheng Liu[1]

1 Fujian Engineering and Research Center of Rural Sewage Treatment and Water Safety, Xiamen University of Technology, Xiamen, China
E-mail: gshliu@qq.com

Abstract. Climate change and economy development are key factors affecting the future of water quality in urbanized catchments. Under the dual influences of climate change and human activities, water environment problems become prominently, especially in coastal watershed of higher urbanization, concentrated populations. Thus, this study focuses on the relationships between economic growth, climate change and water quality in the Jiulong River Watershed (JRW). The results showed that inorganic nitrogen and phosphate nutrient increased significantly, and the river basin showed a trend of warm and humidification. The wind speed and sunshine duration decreased significantly. The results of principal component analysis showed that economic development play a leading role in the evolution of estuary water environment, and the influence of climate change cannot be ignored. Economic growth and climate change contribute to the deterioration of water quality. Water resource shortages and water pollutions are the big challenges for water resources management. This study will be conductive to the water environmental pollution control and management.

1. Introduction

In recent years, under the influences of global climate change and human activities [1], water quality conditions of the estuary area have undergone significant changes. The emergence of water environment problems is the inevitable result of the excessive use of water and soil resources by human beings. Studies by Ahearn (2005) have shown that several changes in the climate system that can have a notable impact on river water quality [2], especially in multi-factor scenarios, including the influence of climatic factors and non-climatic factors. Molina-navarro (2014) showed that climate change had a significant promoting effect on nutrient salt output [3]. Todd et al. (2012) found that the water quality showed a significant trend of deterioration due to the impact of climate change in the Snake River Basin, which is less affected by human interference [4].

The combined effect of rapid urbanization and climate change poses a serious threat to the water environment and water resources security in the JRW. Nutrient salt load in some river and reservoir areas tends to be saturated, and algal bloom has occurred occasionally [5]. Guo (2012) analyzed the production and migration process of major land-based pollutants, including sediment, nitrogen and phosphorus nutrients, heavy metals and organic matter of the JRW, and found that the generation of land-based pollutants was closely related to the water cycle [5].

The dual effects of economic development and climate change on the evolution of water quality have not been considered comprehensively. Therefore, the potential impact of climate change and economic growth on the water environment in a coastal watershed with high urbanization degree, population

Content from this work may be used under the terms of the Creative Commons Attribution 3.0 licence. Any further distribution of this work must maintain attribution to the author(s) and the title of the work, journal citation and DOI.
Published under licence by IOP Publishing Ltd

concentration and high dependence on water resources has been executed. This is crucial to formulate reasonable water resource management and water environment protection.

2. Data and methods

2.1 Study area
Jiulong River is not only the drinking water sources, but also the important water source of industrial and agricultural production of Longyan, Zhangzhou and Xiamen [5]. It is located at subtropical regions, belong to the subtropical ocean monsoon climate. The average annual temperature was between 19.9 and 21.1 °C, the years of average rainfall was 1400-1800 mm.

2.2 Data handling
Meteorological data were obtained from China's meteorological science data sharing service (http:// cdc.cma.gov.cn/). Three meteorological stations are assigned to the JRW, each providing complete daily data on precipitation, air temperature, sunshine duration, and wind speed for the period 1954–2011. Nitrogen and phosphorus nutrient salt is an important pollution index, and also an important factor to regulate and control the comprehensive treatment of water environment in the JRW. The chemical oxygen demand (COD) is an important parameter to evaluate the organic pollution of water body. The historical water quality data of research area are from the researches of Guo (2012) and the Environmental monitoring stations of Fujian province [5]. Some studies also show that the main environmental problems in the estuary are the eutrophication caused by organic pollution and inorganic nitrogen and phosphorus [6].
Therefore, the water quality indexes, inorganic nitrogen (DIN), COD and active phosphate (DIP) in the seawater of the JRW were selected, and the data time series are from 1981 to 2015. In addition, the GDP and other economic development data of Longyan, Zhangzhou and Xiamen from 1980 to 2012 were obtained from the statistical yearbook of Fujian province.

2.3 Methods
The central idea of principal component analysis (PCA) is to reduce the dimensionality of a data set that consists of a large number of interrelated variables, while retaining as many of the variations present in the data set. Numerous domestic and international researchers have used PCA to illustrate the effect of climate factors on the runoff process mechanism [7]. The current paper adopts the same method to determine the main influencing factors and the contributions of such factors to water quality. SPSS 16.0 and Origin 8.0 were used for all the analyses.

3. Results

3.1 Variation tendency
3.1.1 Climate change

Air temperature of the JRW raised significantly (Figure.1, $p < 0.05$) and the growth rate was 0.15 °C/decade, higher than the average level in Fujian province [8]. Precipitation presented a fluctuation increase trend ($p>0.05$), with an increase of 21.4 mm/decade, slightly higher than the average level of Fujian province (28.3 mm/decade). The wind speed presented a significant decreasing trend ($p<0.05$), with a decreasing amplitude of -0.12m /s per decade, slightly lower than the average level of Fujian province (-0.15m/s). Sunshine hours also showed a significant decrease trend ($p<0.05$), with a decrease rate of 80.84h per decade, higher than the national average (-37.6h). Fog, haze and other weather phenomena affecting sunshine hours were increasing day by day, which were the main reasons for the decrease of sunshine hours [8]. On the whole, the JRW showed a trend of warming and humidification, and the wind speed and sunshine hours showed a significant downward trend.

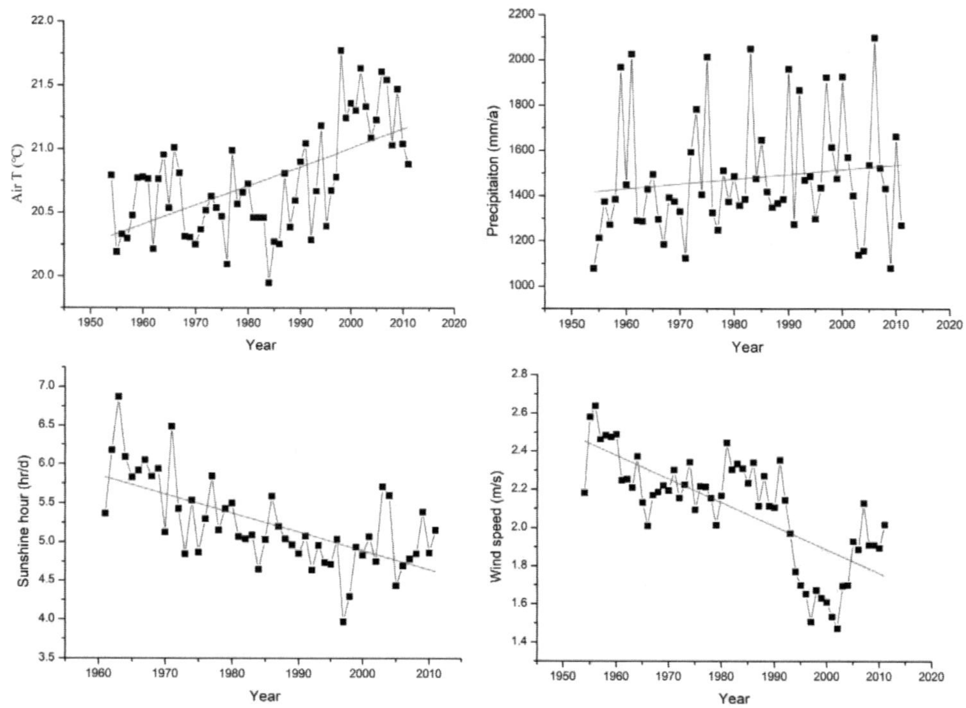

Figure 1. The variation of air temperature, precipitation, wind speed and sunshine hours of the JRW from 1954 to 2011

3.1.2 Economic growth

The GDP of JRW shows a trend of continuous growth, with an overall growth rate of 32.25 billion yuan per year, and the growth rates of primary industry and secondary industry are 1.99 billion yuan per year and 22.084 billion yuan per year respectively (Figure.2). On the whole, the economic indicators of the JRW show an upward trend year by year. The regional GDP continues to grow, and the economy of the basin presents an economic pattern of slow growth of the primary industry and rapid growth of the secondary industry [5].

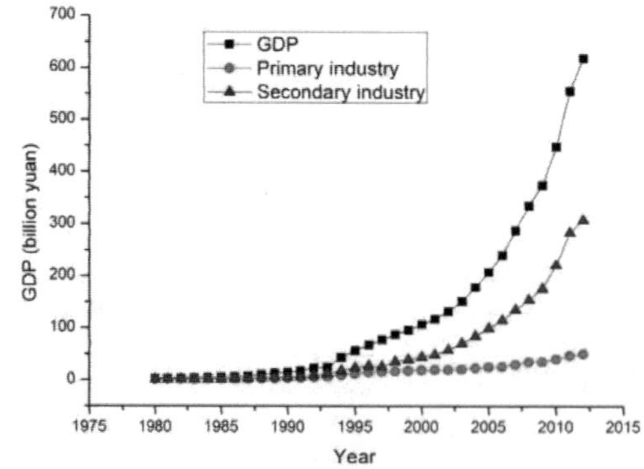

Figure 2. The variation of GDP composition of JRW from 1980 to 2012

3.1.3 Water quality

In the watershed, the DIN and DIP showed a significant increase trend (Figure. 3, $p<0.05$), with an increase of 0.28mg/L and 0.008mg/L per decade, respectively. While the COD showed a "Sin-shaped" insignificant increase trend. In general, the water quality of JRW has been deteriorating in the past 36 years, which is closely related to the application amount of nitrogen and phosphate fertilizer and the rapid development of livestock and poultry breeding industry.

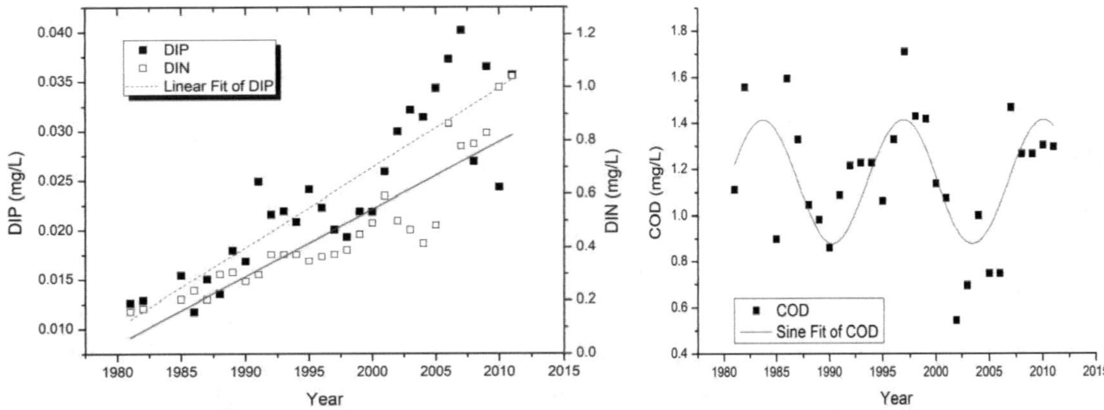

Figure 3. The variation of DIN, COD and DIP of Jiulong River Basin from 1980 to 2015

3.2 Determining the main driving factors of water quality by PCA

The main driving factors of water quality include climate change elements and human activities. The impact of climate factors and economic development on water quality is complex and multi-layered, which is directly or indirectly affected by temperature, precipitation, wind speed, GDP, land use change and other factors. Observed data on air temperature, precipitation, wind speed, sunshine hours, GDP, GDP of the primary industry and the second industry were selected for the PCA test (Tables 1 and 2). Before the rotation of PCA, the Kaiser–Meyer–Olkin (KMO) test was conducted. Obtained KMO values were all greater than 0.6, suggesting that the PCA was highly suitable. The analysis was then used to determine the main driving factors of water quality and the relative contributions of such factors.

Table 1. The principal component rotation load matrix.

Factors	**Principle components**		
	1	**2**	**3**
Air temperature	0.413	**0.772**	-0.034
Rainfall	-0.075	-0.067	**-0.893**
Wind speed	-0.141	**-0.921**	0.102
Sunshine hours	0.039	-0.200	**0.860**
GDP	**0.980**	0.177	0.074
Primary industry	**0.915**	0.392	0.029
Secondary industry	**0.984**	0.149	0.078

Table 2. The principal component contribution rate table.

Principle components	**1**	**2**	**3**
Characteristic value	3.63	1.69	0.91
Contribution rate（%）	51.81	24.13	12.95
Cumulative contribution rate（%）	51.81	75.94	88.89

The accumulated contribution rates of the first three principle components reached 88.9% (Table 1), demonstrating that three main principle components had full generalizations of the seven factors. The first principal components considered in the JRW were GDP, GDP of the primary industry and the secondary industry, with a level of influence on water quality of about 51.81%. This is defined as the economic development principal component, which was affected primarily by economy growth. The second principal component is mainly composed of air temperature and wind speed, which is defined as climatic principal component A. The third principal component is mainly composed of precipitation and sunshine hours, which is defined as climatic principal component B

The contribution rate of the principal component of economic development reached 51.81%, indicating that economic development played a leading role in the water quality change of the basin and was the decisive principal component. The contribution values of climatic principal components A and B are 24.13% and 12.95%, respectively. In general, the total contribution values of climatic principal components also account for 37.08%. Therefore, the impact of climate change cannot be ignored in the evolution of water quality.

4. Discussions

4.1 Impact of economic development and climate change on water quality changes

As shown in Table 3, DIP and DIN showed a significant growth trend with the increase of the main components of economic growth（$p<0.01$, $n=29$）. The Pearson correlation analysis between economic growth component and water quality showed that, the correlation coefficient between DIN and economic development was the largest, exceeding 0.9 ($p<0.05$), and the correlation coefficient by DIP was 0.72 ($p<0.05$). This is closely related to the increase of the planting area and the application of nitrogen and phosphorus fertilizers in the primary industry (Guo, 2012). However, COD has no correlation with economic development ($p>0.05$), indicating that economic development has less impact on COD. The effect of primary industry and secondary industry on the water pollution factor COD is significant in Liaohe River [9]. In general, economic development has a positive feedback effect on the increase of DIP and DIN concentration in the basin.

Table 3. Pearson correlations between water quality and impact factors.

	Economic growth			Climate factors				Principle components		
	GDP	Primary	Secondary	T_a	P	W_s	S_h	Economy	Climatic A	Climatic B
COD	0.022	0.002	0.016	0.228	0.005	0.113	0.137	0.054	0.215	0.009
DIN	0.949*	0.943*	0.941*	0.537*	0.057	0.295	0.039	0.917*	0.005	0.403*
DIP	0.739*	0.797*	0.726*	0.668*	0.111	0.377*	0.015	0.811*	0.046	0.025

Note: Primary is Primary industry, Secondary is Secondary industry, T_a is air temperature, P is precipitation, W_s is wind speed, S_h is sunshine hours. The * in the table indicates a significance level (*p*) of less than 0.05

DIP and DIN concentrations increased significantly with increasing temperature ($p<0.05$). Under the influence of climate change, warming will increase the water temperature of water, pH, and solubility rate of various species. Arheimer (2005) also found that climate warming would increase the nitrogen concentration of the water by 13% and the total phosphorus concentration by 50% [2]. The increase of water temperature will cause a series of reactions in water sediments, promote the dissolution of N, P and other substances. As the temperature rises, the COD concentration in the water body shows a downward trend. This is related to the fact that the temperature rise accelerates the chemical reaction and biodegradation rate in the water, thereby promoting the conversion of water organic matter and decreasing the COD concentration.

As the wind speed decreased, the DIP concentration showed a significant upward trend ($p<0.05$), and the DIN concentration also showed an increasing trend. The decrease of wind speed makes the flow rate of water body slow, which makes the residence time of water body longer, and leads to the increase of N and P concentrations [2]. There was no significant correlation between precipitation and sunshine hours and changes in DIP, DIN and COD concentrations (Table 3). However, with the decrease of climatic component B, the DIN concentration showed a significant downward trend ($p<0.05$. There was a negative correlation between DIN concentration and precipitation, and a positive correlation with sunshine hours.. The increase in precipitation increases the runoff, which determines the degree of dilution of water pollutants [1], which in turn affects the river nitrogen load. Light promotes the release of algae nitrogen and phosphorus in water, which affects the photosynthesis and nitrogen fixation of aquatic plants and promotes the growth of TN in water [8].

4.2 Effects of climatic factor fluctuations on algal bloom

A large-scale outbreak of algal bloom occurred in JRW in 2009. As shown in Figure 4, the outbreak of this algal bloom was extremely likely to occur under the background of temperature increase, precipitation decrease, wind speed decrease and unusually high sunshine hours in January and February 2009. The investigation also found that the quantity of incoming water from the reservoir was relatively small and the temperature was relatively high when this algal bloom event occurred.

Therefore, it can be speculated that under the premise of water eutrophication, these climatic factors play an important role in promoting algal bloom. The role of climate change cannot be ignored in the planning of watershed water environment and water resources. In view of the important impact of algal bloom on regional water environment and water security, corresponding effective measures should be taken to cope with climate change and prevent algal bloom outbreaks.

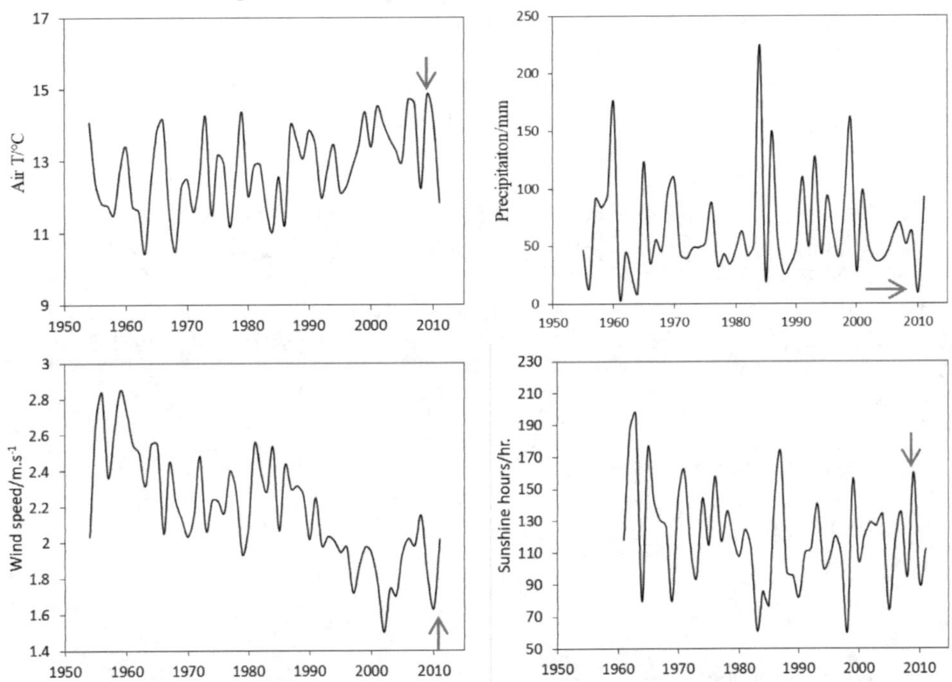

Figure 4. The variation characteristics of average January and February air temperature, precipitation, wind speed, and sunshine hours in JRW

5. Conclusions

The watershed showed a trend of warming and humidification, and the wind speed and sunshine hours decreased significantly, which would promote the algal bloom outbreak. With the continuous economic growth in the JRW, the DIN and DIP increased significantly, and the water quality generally showed a trend of deterioration. The influence of the climatic principal component on the change of COD concentration was more than that of the economic growth principal component. With the increase of temperature and the decrease of wind speed, the DIN and DIP increased significantly, and the COD concentration showed a downward trend.

The contribution rate of the economic development principal components reached 51.8%, indicating that economic development played a leading role in the change of water quality in the JRW. Furthermore, the total contribution value of the climate principal components also reached 37.1%, implying the influence of climate change still cannot be ignored. Climate change has a superposition effect on the deterioration of water quality. Therefore, the impact of climate change should also be paid attention to when the industrial structure is optimized and new environmental protection technologies are applied to improve the water environment quality in the process of water environment pollution.

6. References

[1] Lotstein EL 2013 Global Climate Change Impacts in the United States: A State of Knowledge Report from the U.S. Global Climate Change Research Program *Journal of Environmental Quality* **40(4)** 279

[2] Arheimer B, Andersson J, Fogelberg S 2005 Climate change impact on water quality: Model results from Sourthern Sweden *Ambio Towards More Integrated Water Resource Management*, **34(7)** 559-566

[3] Molina-Navarro E, Trolle D, Martínez-Pérez S 2014 Hydrological and water quality impact assessment of a Mediterranean limno-reservoir under climate change and land use management scenarios *Journal of Hydrology* **509(4)** 354-366

[4] Todd AS, Manning AH, Verplanck PL 2012 Climate-change-driven deterioration of water quality in a mineralized watershed *Environmental Science & Technology* **46(17)** 9324-9332

[5] Guo ZH, Wang C, Yan L 2012 Spatiotemporal variations of main pollutants in Jiulongjiang Estuary *China Environmental Science*, 2012, 32（4）: 679-686.

[6] Zhang L, Cao W, Ying-Qun MA 2016 Distribution of Nitrogen and Phosphorus in the Tidal Reach and Estuary of the Daliao River and Analysis of Potential Eutrophication *Environmental Science* **37(5)** 1677-1684

[7] Yue S & Wang CY 2002 Applicability of prewhitening to eliminate the influence of serial correlation on the Mann-Kendall test *Water Resources Research* **38(6)** 1068-1074

[8] Yang K, Chen BB, Chen JJ 2011 Characteristic of Climate Change in Fujian Province *Chinese Agricultural Science Bulletin* **27(8)** 302-309

[9] Du X, Xu D, Fu X 2015 An empirical research on the relationship between water environment and economic development in liaohe river watershed *ACTA ECOLOGICA SINICA* **35(6)** 1955-1960

Acknowledgments

This research was supported by the Natural Science foundation of China (No. 51809222) and the Open Research Fund Program from Fujian Engineering and Research Center of Rural Sewage Trea tment and Water Safety (RST201809).

IOP Conf. Series: Earth and Environmental Science **356** (2019) 012016 doi:10.1088/1755-1315/356/1/012016

Experimental study on combination of microbial flocculants with PAC in in-situ remediation of black odorous water and sediment

H Y Cheng, J X Long, Z Liu and A L Yang

School of environmental science and engineering, Xiamen University of Technology, 600 Ligong Road, Xiamen, 361024, China
E-mail: 2012111601@xmut.edu.cn

Abstract. This study, the sediment of black and odorous water body in Xinyang main flood drainage channel of Xiamen City was taken as the research object, and the effect of in-situ remediation of black and odorous water body sediment by microbial flocculant (MBF) and polyaluminium chloride (PAC) in different proportions was studied through laboratory static simulation test. The experimental results show that: (a) The MBF dosage of 20 mg/L has a good effect on the remediation of the sediment in the black-odor river. The removal rates of TP and COD in the overlying water are 86.5% and 87.37% after 15 days of experiments, respectively. The removal efficiency of acid volatile sulfur in sediment (AVS) is not very good, the removal rate is only 27.1%. 100 mg/L PAC is added to remedy the black-odor river sediment, and the overlying water under the same experimental conditions. The removal rate of TP was 51%. (b) The combination of MBF and PAC has a good effect on the remediation of the sediment and overlying water of the black and odorous river in the city. The optimum dosage scheme is PAC with 20 mg/L and microbial flocculant with 10 mg/L. Under this scheme, the removal rates of NH_4^+-N, TP and COD in overlying water were 92.37%, 70.82% and 64.52% respectively after 15 days of restoration experiment. AVS in sediment was reduced by 52.21%, which basically eliminated the black odor in overlying water.

1. Introduction

Nowadays, remediation of black and odorous water body is the most popular problem in water environment treatment. The polluted river sediment is the main source of internal pollution which causes black and odorous water body. There are a lot of nutrients in the sediment, which are easy to cause secondary pollution. Therefore, remediation of polluted sediment is particularly important. In-situ chemical remediation technology can effectively curb the problem of black and odor in polluted sediment, and its treatment effect is fast and secondary pollution is small, so it becomes an ideal treatment method. Polyaluminium chloride (PAC), polyacrylamide (PAM) and other inorganic or synthetic organic macromolecule flocculants are commonly used in advanced sewage treatment. However, it will cause insecurity and secondary pollution to the environment in the process of use[1]. Microbial flocculant is a kind of macromolecule substance produced by microbial cells[2]. Its main components are glycoprotein, mucopolysaccharide, protein, cellulose and deoxyribonucleic acid[3-4], etc. This substance has good flocculation performance[5] and is safe and non-toxic. It has attracted wide attention as a green water treatment agent[6]. However, the large dosage and high price of flocculant limit its wide application in large-scale production and industry.

Content from this work may be used under the terms of the Creative Commons Attribution 3.0 licence. Any further distribution of this work must maintain attribution to the author(s) and the title of the work, journal citation and DOI.
Published under licence by IOP Publishing Ltd

2. Materials and methods

2.1. Experimental Materials

The bacteria used in the experiment were separated and preserved from activated sludge of sewage treatment plant. The seed culture medium consists of beef extract, peptone and NaCl. The fermentation medium consisted of sucrose, $(NH_4)\ _2SO_4$, $MgSO_4$, $CaCl_2$, phosphate buffer solution and trace element solution. The inoculation amount of seed liquid was 3% (V/V). After 72 hours of fermentation, 50mL fermentation broth was centrifuged. The supernatant after centrifugation was mixed with absolute ethanol at a ratio of 1:3. The supernatant was centrifuged for 16 hours, 5000r per minute for 15 minutes. The supernatant was poured out and dried to prepare microbial flocculant solution. Polyaluminium chloride (PAC) is prepared from solid polyaluminium chloride. Other concentrations of microbial flocculant and polyaluminium chloride flocculant used in the experiment are obtained by diluting the concentration of PAC.

2.2. Experimental Instruments and Methods

The instruments used in the experiment include: ultraviolet-visible spectrophotometer, electronic analytical balance, oven, digestion instrument, centrifuge, etc. The indexes of overlying water include ammonia nitrogen (NH_4^+-N), total phosphorus (TP), and biochemical oxygen demand (COD). The methods of detecting these indexes are consulted with relevant literatures[7]. The reason why sludge organic sediment shows black odor is that there are a lot of sulfides in the sediment. The device for measuring AVS is shown in the Figure 1.Principle of AVS [8]determination reaction: adding a certain concentration of hydrochloric acid and blowing with nitrogen to escape sulfide from sediment in the form of H_2S gas, and then absorbed by NaOH solution, colorimetric analysis of the amount of H_2S absorbed by NaOH was carried out.

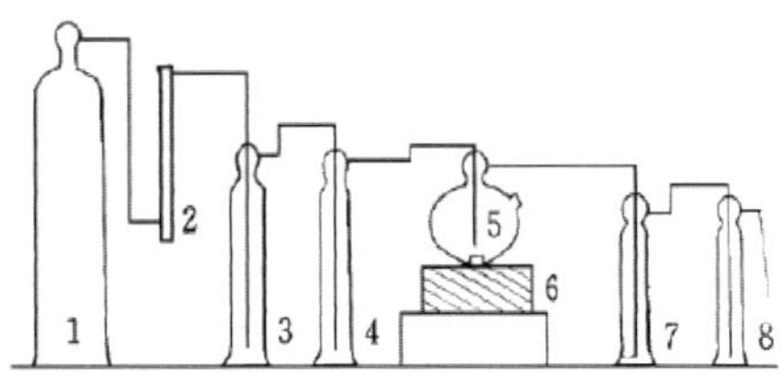

Figure 1. Sketch of AVS Measuring Reactor.

(1. Nitrogen cylinder. 2. Gas flow meter

3-4. Gas cylinder. 5. Flat-bottomed Reaction Bottle. 6. magnetic stirrer. 7-8. Gas absorption bottle)

3. Result and analysis

3.1. Effect of PAC on Sediment Rehabilitation and Overburden Water

The concentration of 100 mg/L polyaluminium chloride (PAC) was added into the bottom mud, the solution of polyaluminium chloride was injected into the bottom mud, distilled water was added into the bottom mud, and placed in a 26-degree thermostatic incubator. The dosage was as follows: blank control test No. 1, 100 mg/L wet mud was added to No. 2, and the measuring days were 1, 3, 5, 7, 9, 12 and 15 days. The change of AVS in the bottom mud was measured, and ammonia nitrogen (NH_4 nitrogen) in the overlying water was also detected.+ - N, total phosphorus (TP), biochemical oxygen demand (COD) in water. The effects of PVA as a restorative agent on sediment AVS and NH_4^+-N, TP, COD in overlying water were shown in Figure 2 to Figure 5, respectively.

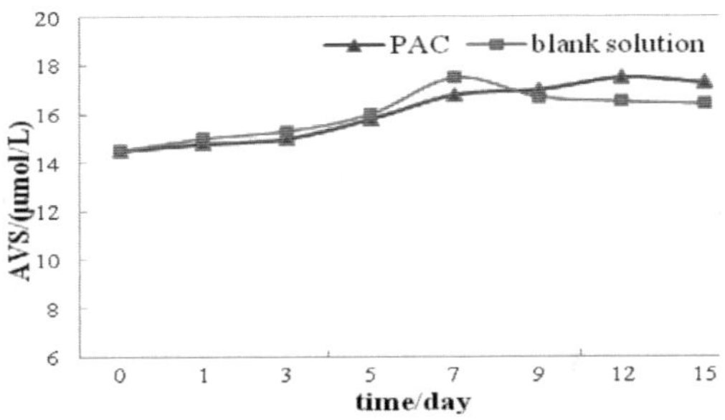

Figure 2. Effect of PAC on AVS in sediment.

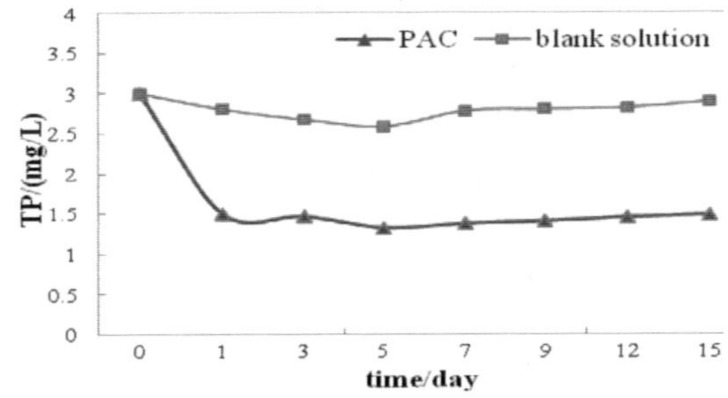

Figure3. Effect of PAC on TP in overlying water.

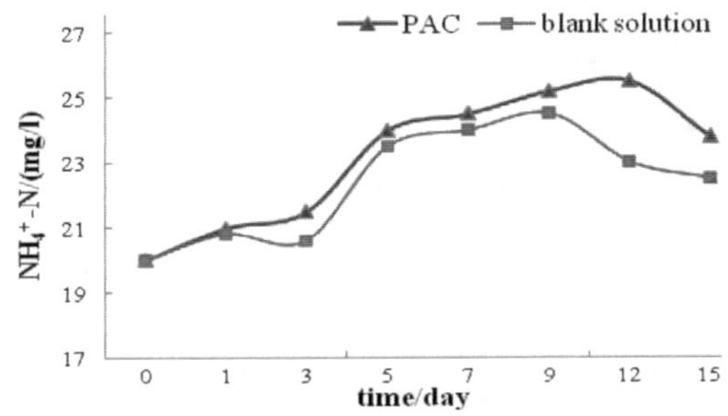

Figure 4. Effect of PAC on NH_4^+-N in overlying water.

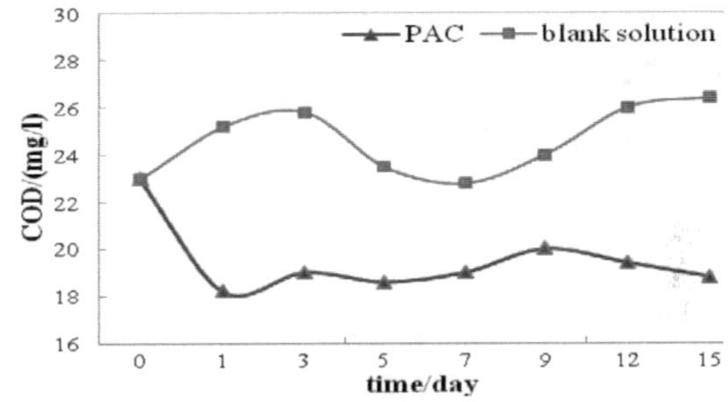

Figure 5. Effect of PAC on COD in overlying water.

As shown in Figure 2, when PAC is added to the bottom mud, the removal effect of AVS pollution of the bottom mud is not obvious, but remains at a level. AVS of the bottom mud remains at 14-18 umol/L. Compared with the blank control group, the floatation is smaller and there is a small increase. Polyaluminium chloride will form flocculent layer on the surface of the bottom mud.

While in Figure 3, NH_4^+-N in sediment increased to a certain extent, increased slightly in a short time, followed by a downward trend. PAC hydrolyzed on the surface of sediment to form $Al_2(OH)_3$ flocculant, and formed a cover layer at the interface of mud and water. From 15 days' observation, it can be found that $NH4^+$-N in overlying water increased slightly in a short time, but the increase was not obvious.

From Figure 4, the inhibition of PAC on phosphorus was very obvious in the early stage of reaction, especially in the first five days. Compared with the blank reactor, the TP content in the overlying water was significantly reduced and stabilized at the level of 1.5 mg/L. The blank control group had no effect because of the absence of it.

As can be seen in Figure 5, after PAC was injected into sediment, COD in overlying water decreased significantly and remained at a low level, at 18-20 mg/L. The possible reason[9] is that aluminium salts are hydrolyzed to form Al^{3+}, and aluminium hydroxide produced by Al^{3+} adsorbs humic acid, humus and other macromolecule organic matter in sediment. Small molecule organic matter will be adsorbed by the adsorption sites in the flocculent layer of aluminium hydroxide when it diffuses to the overlying water, so adding PAC will reduce the COD in overlying water.

3.2. Effect of MBF on Sediment Rehabilitation and Overlying Water

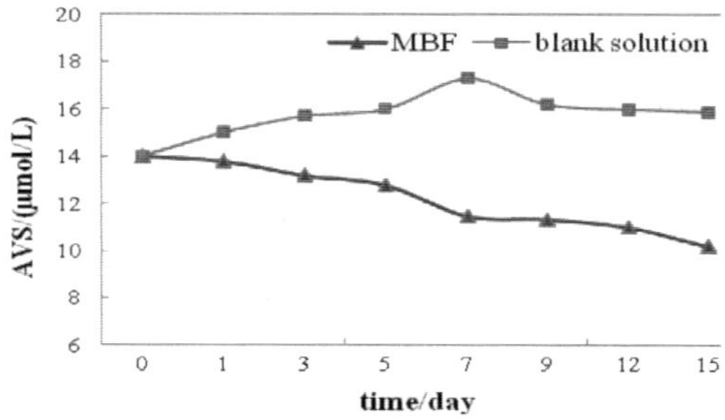

Figure 6. Effect of MBF on AVS in sediment.

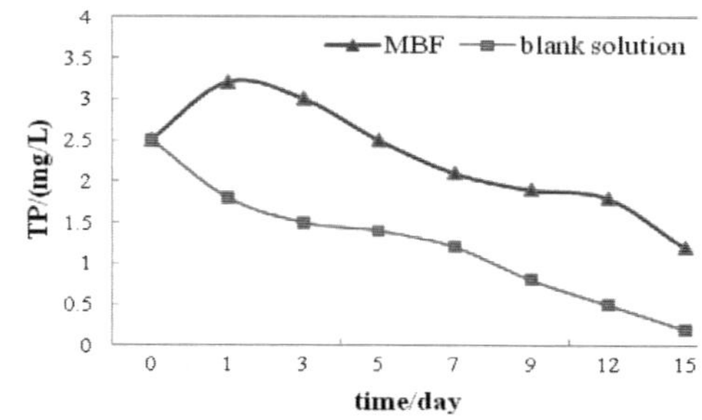

Figure 7. Effect of MBF on TP in overlying water

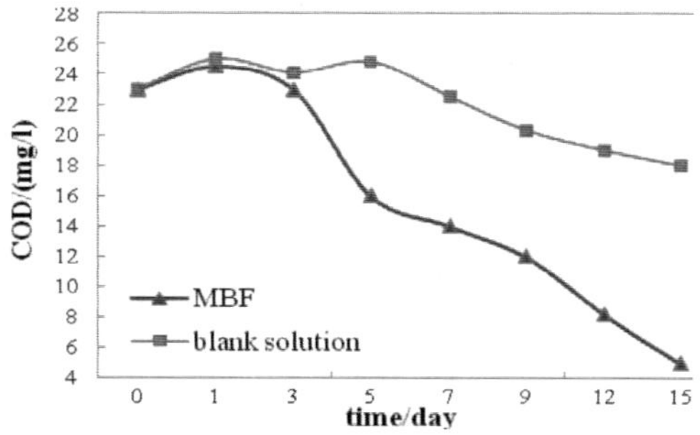

Figure 8. Effect of MBF on COD in overlying water.

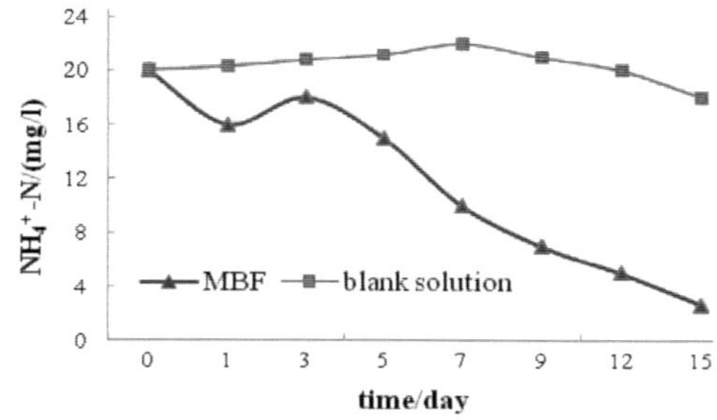

Figure 9. Effect of MBF on NH_4^+-N in overlying water.

It can be seen from Figure 6 to Figure 9 that MBF has a good effect on the removal of NH_4^+-N, COD and TP remove. Even the removal rate of AVS from sludge is higher, which is similar to the results of relevant researchers [10].

3.3. Effect of PAC and MBF on Sediment Rehabilitation and Overlying Water

In order to reduce dosage and enhance flocculation effect, the flocculation effect of treating secondary effluent by mixing PAC and MBF according to different dosage was investigated.

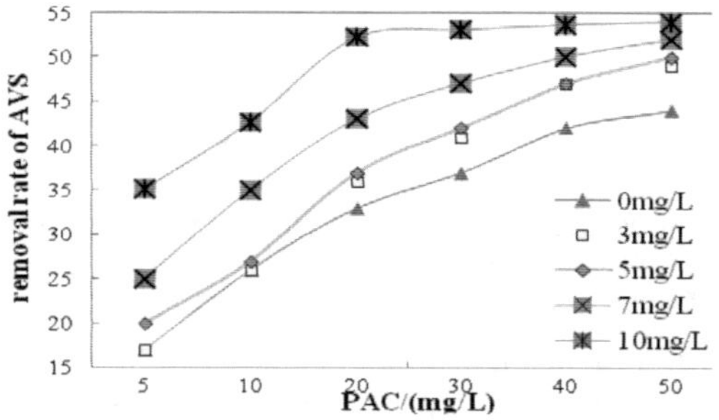

Figure 10. Effect of combination of PAC and MBF on AVS in sediment

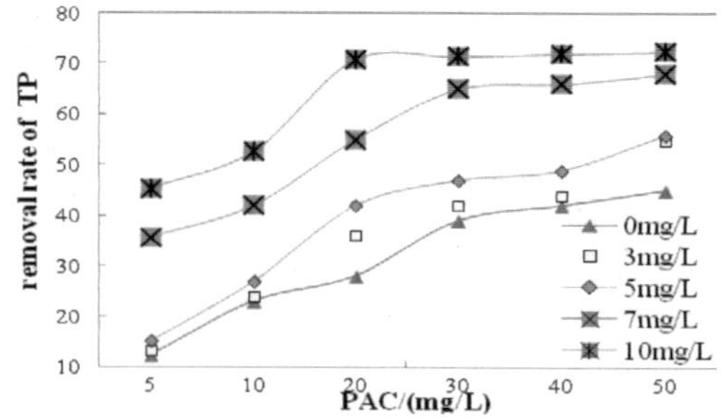

Figure 11. Effect of combination of PAC and MBF on TP in overlying water.

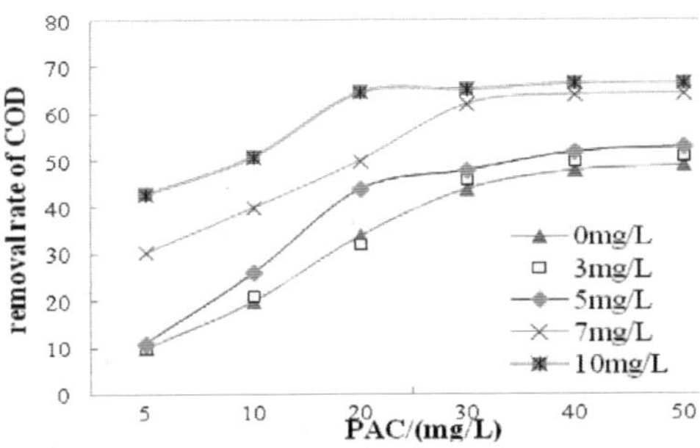

Figure 12. Effect of combination of PAC and MBF on COD in overlying water.

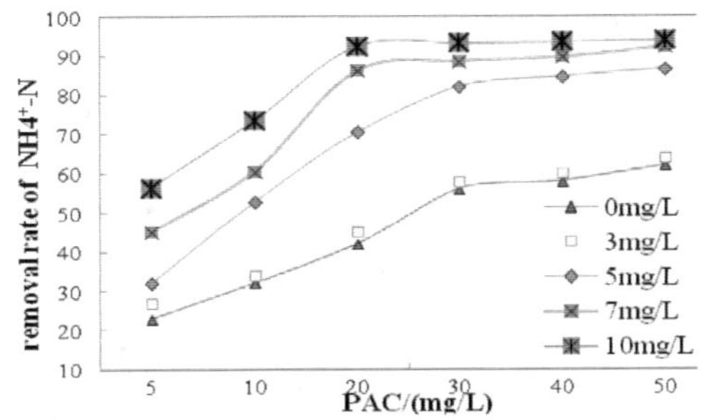

Figure 13. Effect of combination of PAC and MBF on NH4+-N in overlying water.

It can be seen from the Figure 10 to Figure 13 that the removal rate of AVS, NH_4^+-N, COD and TP can be significantly increased by the combination of microbial flocculant and PAC. The removal rate increases with the increase of PAC and MBF dosage. This is because on the one hand[11], the addition of cations neutralizes the negative charges on the functional groups of microbial flocculants and colloidal particles, which makes it easier to bridge with microbial flocculants. On the other hand, through the adsorption and coagulation of cationic flocculants, colloidal particles can be more effectively adsorbed and netted by microbial flocculants [12]. When PAC dosage was 20 mg/L and MBF dosage was 50 mg/L, COD removal rate and TP removal rate reached 64.52% and 70.82%, respectively. When 20 mg/L PAC was added alone, COD removal rate and TP removal rate were 18.2% and 51%, respectively. When MBF dosage was 50 mg/L, when PAC dosage was more than 20 mg/L, the removal rates of COD and TP did not increase significantly. Therefore, microbial flocculant 50 mg/L and PAC 20 mg/L were selected as the optimal dosage.

4. conclusion

(1) When MBF is used alone to treat overlying water, when the dosage range is 10-20 mg/L, MBF can effectively remove COD and TP from the water, but the excessive dosage will reduce the flocculation efficiency and affect the quality of the effluent. The optimal dosage is 20 mg/L.
(2) When the overlying water is treated by PAC alone, the coagulation effect of AVS is worse than that of MBF alone, and that of TP is the best. MBF has stronger humus removal ability than PAC, but its microbial metabolites removal ability is not strong.
(3) The combination of MBF and PAC can reduce 50% of MBF dosage and 33% of PAC dosage. The optimal dosage is 50 mg/L MBF and mg/L PAC. At this time, COD removal rate reaches 64.52% and TP removal rate reaches 70.82%. Among the components of secondary effluent, the removal of fulvic acid by compound coagulant is the most significant, followed by humic acid.

5. References

[1] Ma F, Li DP, Zheng LN. Treatment of High Algae Water by compound bioflocculant and polyaluminium ferric chloride 2008 J. *China Water and Wastewater*, 24 39
[2] Jin HX, Ma F, Meng L. Compound of microbial flocculant and chemical flocculant and Its application 2006 *J. Progress in Chemical Engineering*, 25 105
[3] Zhu YB, Ma F, Yang JX. Study on compound flocculant and compound flocculant 2010 *J. Journal of Harbin Institute of Technology*, 42 1254
[4] SALEHIZADEH H, YAN N. Recent advances in extracellular biopolymer flocculants 2014 *J . Biotechnology Advances*,32 1506
[5] Mao J, Nie MQ, WU Y. Study on removal of SS from water by mixed microbial flocculants and inorganic ions 2008 *J. Water treatment technology*, 34 11
[6] Meng F, Yang JX, Ma F. Influencing factors of low temperature and low turbidity water treatment with composite bioflocculant 2009 *J. Journal of Harbin Institute of Technology*, **41** 42

[7] 2017 *Water and Wastewater Monitoring and Analysis Method of State Environmental Protection* Administration*(4th Edition).* (Beijing: China Env. Sci. Publishing House)
[8] Lin YH, Guo MX, Zhuang Y. Determination of acid volatile sulfur in sediment and synchronous leaching metal 1997 *J. Journal of Environmental Science*, 3 97
[9] Nwodd UU, Green E, Mabinya LV. Bioflocculant production by a consortium of Streptomyces and Cellulomonas species and media optimization via surface response model 2014 *J. Colloids and Surfaces B: Biointerfaces*,116 257
[10] Li KM, Liu J, Jiang D. Bioremediation and maintenance of Guliaoyong Black Odor Water 2005 *J. Journal of Applied and Environmental Biology*, 6 742
[11] Morett T, Yadav JS, Yan S. Extracellular polymeric substances of bacteria and their potential environmental applications 2014 *J. Journal of Environmental Management,* **144** 1
[12] Xu ZX, Zhang JP, Liao ZL. Effect of Suzhou River sediment on water quality pollution of overburden water 2005 *J. Urban Environment and Urban Ecology*, **6** 1

Acknowledgments:
This research was supported by funding from XMUT with Grant EBL2018005, Key Laboratory of Environmental Biotechnology, Fujian Province University.

Study on water quality of Xiamen sea area

Y J He[1] and Z N Dai[1,2*]

1 School of Environmental Science and Engineering, Xiamen University of Technology, China
2 Key Laboratory of Environmental Biotechnology（XMUT），Fujian Province University
E-mail: daizn@xmut.edu.cn

Abstract. th the rapid development of the economy, it has brought huge pollution load pressure to the marine environment quality. The land-based pollutants have increased year by year, and the marine pollution problem has become more and more serious. The governance of the marine environment has become a serious problem. This paper investigated the water quality of Xiamen. The results showed that the average concentration of heavy metals was 2.16×10^{-3} mg/L for copper element, 2.84×10^{-5} mg/L for lead element, 7.61×10^{-5} mg/L for cadmium element, 1.97×10^{-3} mg/L for arsenic element and 3.66×10^{-3} mg/L for zinc element. The monitoring concentration of mercury element at each monitoring point is lower than 1×10^{-5}mg/L. The main pollutants in the general indicators are active phosphate and inorganic nitrogen. This paper can provide data support for the future management of marine environmental problems in Xiamen Government, and it has positive significance for its future governance.

1 Introduction

China is a large ocean country with 32,000 kilometers of ocean lines and abundant marine resources. Since the 1980s, China's marine economy has increased year by year and has gradually occupied a large part of GDP [1]. However, with the continuous improvement of the marine economy, varieties of marine environmental pollution problems have followed it. In 2017, various types of marine disasters caused a total economic loss of 6 billion yuan. Among them, a total of 68 times red tide disasters caused by marine environmental problems were found, with a total area of 3,679 square kilometers[2]. while the problem of marine pollution has increased year by year, China has also continuously strengthened its work on marine environmental protection. However, the issue of marine environment is still an urgent and big problem. The rapid development of society has destroyed the original ecological balance between human and nature. A large amount of nutrients such as nitrogen and phosphorus enter the waters of rivers, lakes, and reservoirs. Under the condition of sufficient nutrients, algae and other planktons multiply and water quality deteriorates [3]. The heavy metal pollution in the water mainly comes from industrial chemical enterprises or pesticides. It exists in various forms, mainly invading into the human body through diet, which seriously affects its health and even causes metal poisoning [4]. The heavy metal pollution in the water mainly consists of suspended solids and sediments, which cause pollution to water bodies. The recovery period is long and difficult, and the pollution may bio-accumulated in marine organisms and cause damage to them [5].

Xiamen City occupies an important strategic position in the social and economic development of China. With the continuous development of the marine economy, the various marine pollution problems that accompany it have gradually attracted people's attention. Therefore, the indicators of seawater in the coastal waters of the city were investigated, and the pollution status of conventional indicators and heavy metal pollutants in seawater were studied. The purpose of this research is to

Content from this work may be used under the terms of the Creative Commons Attribution 3.0 licence. Any further distribution of this work must maintain attribution to the author(s) and the title of the work, journal citation and DOI.
Published under licence by IOP Publishing Ltd

investigate and study the water pollution of Xiamen's sea area. It can provide scientific information for the governance of marine environmental.

2 Materials and methods

2.1The arrangement and sampling of the monitoring point

The water samples were collected from seven sites of Xiamen City, as shown in Figure 1, labeled 1 to 7. The total number of samples is 156. The water samples were collected from the surface layer (0-10m), the middle layer (10-20m) and the bottom layer (20-30m). If the seawater depth is less than 20m or 10m, only the surface layer and the middle layer or surface layer was sampled. The sampling time was March, July and November in 2017.

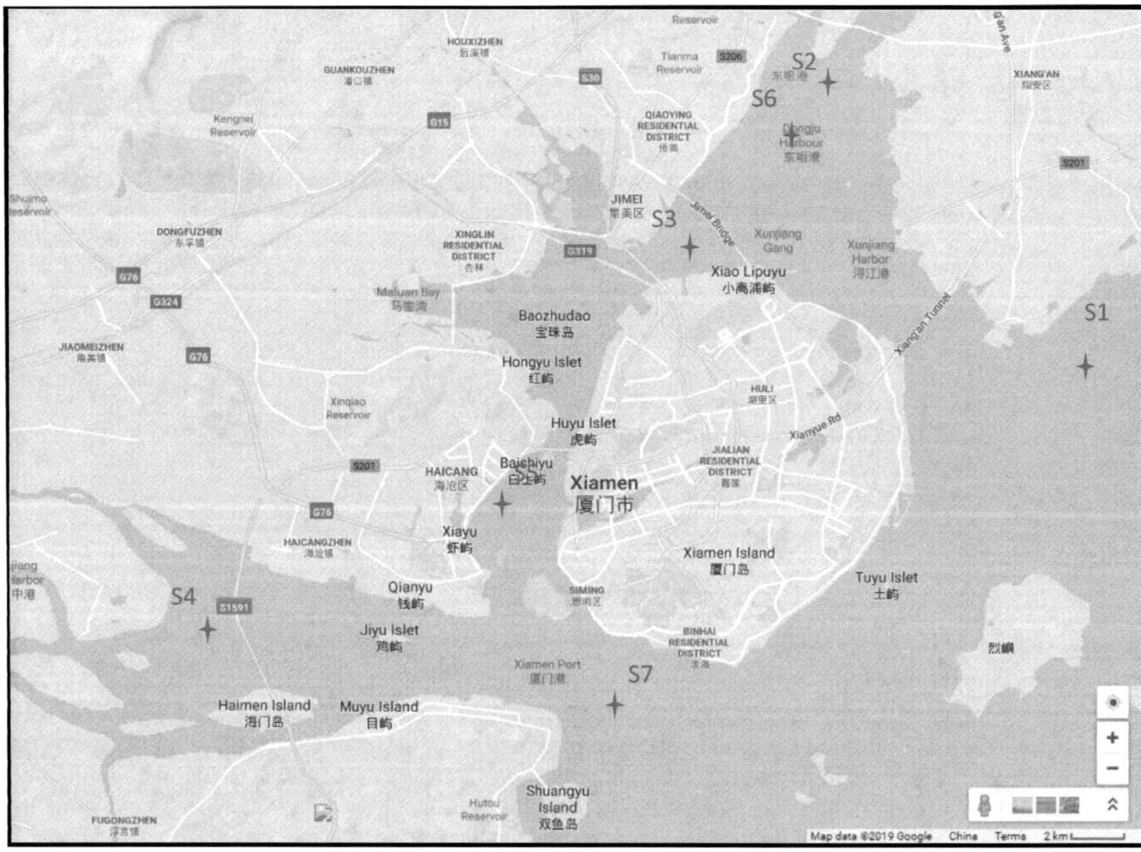

Figure 1. Sampling sites in Xiamen sea area

2.2 Sample Analysis

The water sample is stored in a polyethylene bottle and seal back to the laboratory. The sample is first filtered with a 0.45 μm filter, and the filtered sample is acidified by adding 1% pure nitric acid, followed by heavy metal analysis. The instrument Agilent 7500ce ICP-MS was used for heavy metal analysis to analyze six elements of mercury, copper, lead, cadmium, zinc and arsenic in the sample.

The method of analysis was carried out according to the standard method of EPA 200.8 in the United States, and the standard materials for analysis were all Agilent's multi-metal mixed standard solutions. After each standard curve was completed, the accuracy of the experiment was checked with the domestic environmental standard sample GSBS 5009-88 (200928) to ensure that the measured values of the samples were within the true value.

3 Results and discussion

3.1 General indicators

The monitoring time is March, July and November, corresponding to the three periods ——wet season, level period and dry season of the Xiamen sea area. General indicators were determined for water samples from seven monitoring points. The water quality of the coastal waters in Xiamen is listed in Table 1.

Table 1. Water quality status at seven monitoring points (the unit is mg/L).

Monitoring points	Water temperature ℃	salinity Cl‰	Suspended matter (mg/L)	Dissolved oxygen (mg/L)	pH	Active phosphate (mg/L)	COD (mg/L)	Inorganic nitrogen (mg/L)
S1	25.16±3.13	30.90±2.15	24.90±17.65	7.16±1.24	8.09±0.07	0.02±0.03	0.62±0.34	0.35±0.03
S2	23.25±6.89	30.03±1.48	17.39±8.63	7.50±1.76	8.03±0.17	0.04±0.03	0.77±0.46	0.57±0.02
S3	22.71±8.20	28.30±1.64	15.33±7.34	8.14±3.15	7.85±0.40	0.06±0.06	1.09±0.40	0.66±0.04
S4	24.25±1.63	26.93±1.29	16.17±8.92	7.28±1.44	7.97±0.15	0.03±0.02	0.94±0.36	0.76±0.03
S5	25.79±3.18	27.27±1.13	15.97±9.38	7.24±1.43	8.11±0.23	0.04±0.03	0.99±0.45	0.77±0.03
S6	22.71±8.13	30.08±1.28	17.67±8.62	7.56±1.22	7.92±0.29	0.04±0.03	0.97±0.18	0.48±0.03
S7	24.26±1.65	26.93±1.30	15.17±9.61	7.38±1.28	7.98±0.11	0.03±0.02	0.94±0.31	0.75±0.03

It can be seen from the table that the pH range is 7.92-8.09, dissolved oxygen (DO) concentration range is 7.16-8.14 mg/L, and chemical oxygen demand (COD) is in the range of 0.62-1.09 mg/L, each site meets the national first-class water quality standards. The active phosphate concentration range is 0.02-0.06mg/L, and the sample sites 1, 4, and 7 meet the national second-class water quality standard. The concentration of inorganic nitrogen ranges from 0.35 to 0.77 mg/L, which exceeds the national water quality standards[6].

The above data shows that the excess substance in the coastal waters of Xiamen is mainly active phosphate and inorganic nitrogen. The sampling sites 1, 4 and 7 are relatively better. The reason may be the location of other sites that they were in the middle sea areas in Xiamen Island and Xiang'an District, Jimei District and Haicang District, and the superposition of land source inputs on both sides increases the pollutants discharged into the water body and causes an increase in the concentration of inorganic nitrogen and active phosphate.

Comparing the data of Xiamen conventional indicators in Wu's study[7] in 2016, the concentration of DO, active phosphate, COD and inorganic nitrogen increased; but pH decreased. The increase of active phosphate, COD and inorganic nitrogen concentration, has a great impact on water quality, which will cause pollution such as red tide and affect marine ecology and fishery.

Table 2. Water quality in coastal waters of different cities in 2017 [8-10]

City	pH	DO (mg/L)	Active phosphate (mg/L)	COD (mg/L)	Salinity Cl‰	Tempe-rature ℃	Suspended matter (mg/L)	Inorganic nitrogen (mg/L)
Xiamen	7.99	7.46	0.0360	0.902	28.63	24.02	17.51	0.620
Quanzhou	8.16	7.29	0.0165	0.598	32.68	23.47	23.607	0.193
Putian	8.14	7.41	0.0165	0.611	31.75	22.2	32.66	0.216
Ningde	7.85	7.57	0.0258	0.589	29.208	22.59	40.926	0.306
Fuzhou	8.14	7.68	0.0138	0.617	29.611	22.44	28.198	0.197

The data in Table 2 shows that the DO and COD in five cities in Fujian are all in line with the national first-class water quality standards. However, the active phosphate concentrations, inorganic nitrogen and COD in Xiamen is higher than the other four cities. Years of monitoring also showed that the main pollution in coastal waters in Xiamen is still inorganic nitrogen and active phosphate. It may be affected by the water quality of the Jiulong River. A large number of pollution sources in the Jiulong River Basin continue to bring total nitrogen and total phosphorus to Xiamen's waters, with contributions of about 70% and 30%. In addition, Xiamen sea area and Tongan Bay as semi-enclosed bays have relatively weak self-purification capacity, which is one of the reasons for the further deterioration of water pollution in Xiamen's coastal waters [11].Comparing with Rincon[12] which in Puerto Rico, Xiamen's DO is higher than Rincon. In contrast to the Arabian Gulf[13], Xiamen's DO are slightly higher than it, and active phosphate are lower than the Arabian Gulf. It indicated that the water

IOP Conf. Series: Earth and Environmental Science **356** (2019) 012017 doi:10.1088/1755-1315/356/1/012017

quality in Xiamen is relatively good, getting better year by year.

3.2 Heavy metal

Heavy metals are mostly non-degradable toxic substances, and they do not have any natural purification ability. If they flow into the ecological environment, they will be difficult to eliminate from the ecological environment [14]. The heavy metals studied in this paper are mercury, copper, lead, cadmium, arsenic and zinc. However, the content of mercury is far lower than the national standard of seawater standards and the impact of time and space is negligible. Therefore, the mercury element data is not list in the table. The average concentration of the other five elements during the measurement is shows in Figure 2.

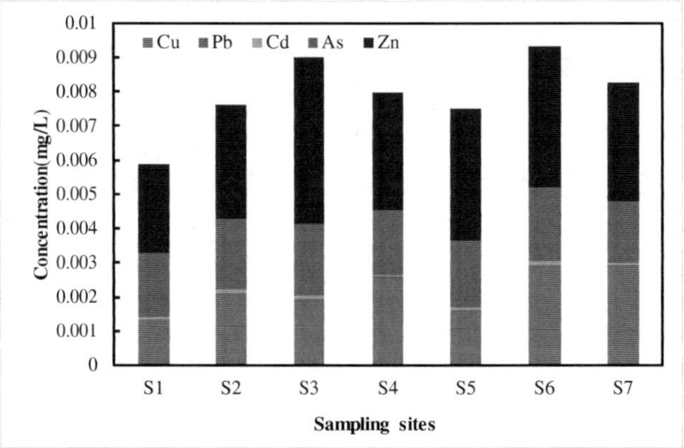

Figure 2. Average concentration of heavy metals in Xiamen sea area.

From the data in Table 3, the concentration of copper is in the range of 1.31-2.9 $\times10^{-3}$mg/L, lead is in the range of 3.93-8.08$\times10^{-5}$ mg/L, and cadmium is in the range of 5.48-9.67$\times10^{-5}$ mg/L. The concentration of arsenic is in the range of 1.78-2.1610^{-3}mg/L, and the concentration of zinc is in the range of 2.57-4.85$\times10^{-5}$ mg/L. The concentration of these heavy metals is lower than the national first-class water quality standards, and it cannot constitute harm to the environment. As can be seen from the figure, the heavy metal content of S3 and S6 is higher than that of other sites, and it is speculated that the S3 and S6 have industrial wastewater discharge.

When Compares with Quanzhou, Ningde, Putian and Fuzhou [8-10], the concentration of copper is higher in Quanzhou and lower in Fuzhou. The concentration of lead is highest in Ningde and lowest in Xiamen. The concentration of cadmium is highest in Putian and lowest in Xiamen. For the concentrations of zinc, Fuzhou is the highest and Quanzhou is the lowest. In general, Xiamen's heavy metal concentration is relatively good than other city. Comparing with the Rigs-to-Reefs region[15], the concentration of these five heavy metals in the Rigs-to-Reefs region was higher than that of Xiamen——the copper concentration is 2.5$\times10^{-3}$ mg/L, the lead element concentration is 3.4$\times10^{-4}$ mg/L, the cadmium element concentration is 4.35$\times10^{-4}$ mg/L, the arsenic element concentration is 2.61$\times10^{-3}$ mg/L, and the zinc element concentration is 7.4$\times10^{-3}$ mg/L.

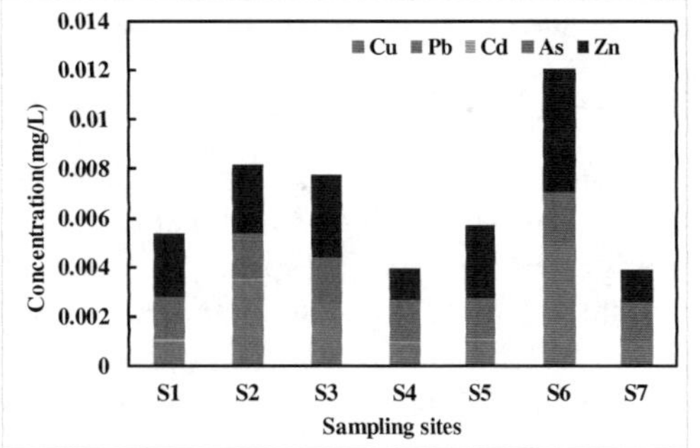

(a)

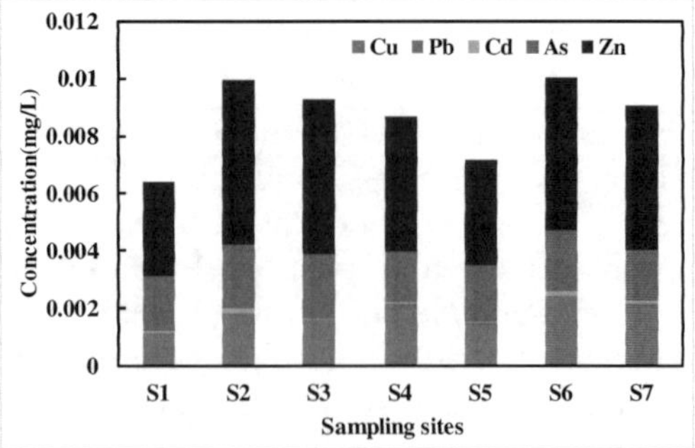

(b)

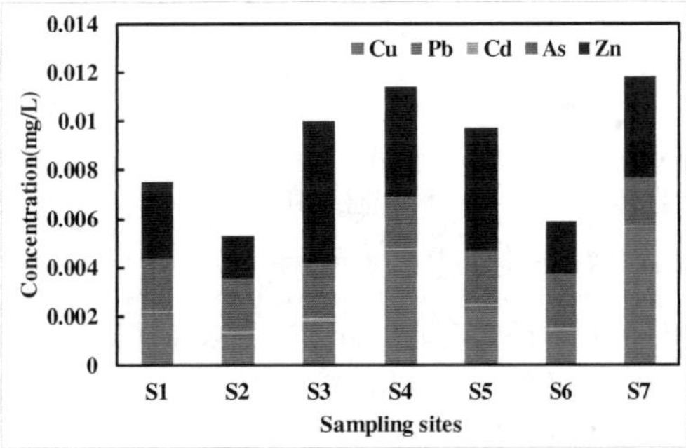

(c)

Figure 3. Concentration of heavy metals in wet season (a), level period (b) and dry season (c) in 2017.
As shown in figure 3, zinc, arsenic and copper account for a large proportion of heavy metals in each period, while lead and cadmium account for a small proportion. The comparison of the three periods shows that the concentration of lead, cadmium and arsenic is dry season > level period > wet season; the concentration of copper is dry season > wet season > level period; the concentration of zinc is level period > dry season > wet season. The common point is that the concentration in the dry season is greater than that in the wet season. The presumed reason is that the heavy metal pollutants discharged into the water body are almost the same in each period, but only the rainwater in the wet season leads to
a decrease in concentration.

4 Summary

In general, the concentration of heavy metal pollutants in the surface seawater of the coastal waters of Xiamen is far lower than the requirements of the national first-class seawater standards, and it cannot pose certain harm to the environment. Lead, cadmium and arsenic have great similarities in time distribution and spatial distribution, and seasonal and spatial changes have negligible effects. Lead and cadmium concentrations are mostly below 1×10^{-4}mg/L, the content of arsenic relative to lead and cadmium is high, and the annual average concentration is 195×10^{-5}mg/L, but it is still within the safe range.

5. References

[1] State Oceanic Administration. "Chinese Marine Economic Statistics Bulletin of 2018" [EB/OL]. [2019-04-11] http://gi.mnr.gov.cn/201904/t20190411_2404774.html

[2] State Oceanic Administration. The State Oceanic Administration issued the "2016 China Marine Environmental Quality Bulletin" [J]. Marine Environmental Science, 2018, 20(2): 75.

[3] Jiang J H. Technical analysis of aquatic plants used in the prevention and control of eutrophication of water bodies [J]. Green Science and Technology, 2019 (04): 96-98.

[4] Zhang J T, Liu J L. The harm of water metal pollution in water and its treatment [J]. Shandong Industrial Technology, 2019 (08): 35.

[5] Meng D, Zhou L Z, Yu C W. Current Status and Treatment Technology of Heavy Metal Pollution in Water[J].Liaoning Chemical Industry,2006(09):534-536+549.

[6] National Environmental Protection Agency. Standard for Marine Water Quality of the People's Republic of China (GB3097-1997) [S]. Beijing: China Standard Press, 1997.

[7] Wu J Q. Comparative Study on Water Quality in Coastal Waters of Different Provinces[J]. Journal of Wuyi University, 2017, 36(09): 28-34.

[8] Fujian provincial department of ecological environment. 2017 Water quality monitoring information of coastal waters in the coastal waters. [EB/OL]. [2017-06-30] http://hbt.fujian.gov.cn/zwgk/kjjc/hjzl /jahysz/201708/t20170822_104032.htm.

[9] Fujian provincial department of ecological environment. 2017 open water quality monitoring information for coastal waters. [EB/OL]. [2017-09-20] http://hbt.fujian.gov.cn/zwgk/kjjc/hjzl /jahysz/201801/t20180123_2040155.htm.

[10] Fujian provincial department of ecological environment. In 2017, the water quality monitoring information of the coastal waters in the dry season was published. [EB/OL]. [2018-01-23] http://hbt.fujian.gov.cn/zwgk/kjjc/hjzl/jahysz /201801/t20180123_2040161.htm.
[11] GUAN Y Z. Research on current situation and countermeasures of water pollution in coastal waters of Xiamen City[J]. Ocean Development and Management, 2007(01): 136-138.
[12] Norat R J, Méndez L P, E.A. Hernández D E A, H. Mattei T L. Cordero R L. A septic waste index model to measure the impact of septic tanks on coastal water quality and coral reef communities in Rincon, Puerto Rico[J]. Ocean and Coastal Management, 2018(12):201-213.
[13] Mostafa A M, Mohammed N, Sami A. Phytoplankton abundance in relation to the quality of the coastal water – Arabian Gulf, Saudi Arabia[J]. Egyptian Journal of Aquatic Research, 2017(10):275-282.
[14] Lu H, Peng Q. Status and Treatment Technology of Heavy Metal Pollution in Water[J]. Contemporary Chemical Engineering, 2019(01): 22-23.
[15] Zheng Y, Zheng X, Gao Z, Zhang Y. Prediction of seawater quality in Rigs-to-Reefs a rea based on grey systems theory[J]. 2013 International Symposium on Environmental S cience and Technology (2013 ISEST), 2013(4):236-242.

Acknowledgments

This study was supported by the Key Laboratory of Environmental Biotechnology（XMUT）, Fujian Province University (EBL2018003) and Xiamen science and technology planning project (3502Z20183064).

Study on water purification ability and permeability of different filter materials in LID-type ditch

M LZhu[1] and Y Chen[1]

1 Water Resources and Environmental Institute, Xiamen University of Technology, Xiamen, CHN
E-mail: 292890310@qq.com

Abstract. In rural areas of China, sewage is often discharged to ditch directly without any treatment. We propose a new type of ditch based on Low Impact Development (LID) strategy by replacing bed soil in rural ditch with mixed filter material of approximately 30cm in thickness. This new type of ditch is so-called LID-type ditch having the function of storage,penetration, purification and drainage. Four basic materials, red loam, sand, gravel and zeolite, are used to create 3 types of mixed filter materials by evenly mixing these basic materials with different mass ratios. Their saturated permeability coefficients and water purification ability are measured and compared. Results indicate that all 3 types of mixed filter materials are suitable to apply to LID-type ditch, and the all outflows meet the water quality standard for farmland irrigation. The best mixed filter material is type 1 with 50% of red loam, 30% of sand, 10% of gravel, and 10% of zeolite, it has removal rate for TP as high as 81.31% and average removal rate of 47.09% for TP, TN, NH3-N, COD, BOD_5.

1. Introduction

In the United States of America, traditional decentralized sewage treatment systems usually consist of two components: a buried septic tank and a soil dispersal system. Figure 1 is a schematic diagram of the decentralized treatment system of rural sewage in the United States [1]. After the treatment of septic tanks in rural areas in the United States, outflow almost becomes clear water and meets the secondary treatment standards stipulated by "the US Federal Water Pollution Prevention and Control Act", and then outflow is discharged and infiltrated into the local soil. The secondary treatment standards are: 7-day averages of BOD_5 and TSS≤45mg/L, 30-day average of BOD_5 and TSS≤45mg/L, 30-day average of pH needs to be in the range of 6-9 [2].

In China, strengthening the construction of rural sewage treatment projects and improving the effectiveness of treatment have been stressed by the government work report. However, compared with the American current situation of rural sewage decentralized treatment, China currently has the following problems: 1) the sewage treatment systems in many rural areas simply copy the first-level-A standard of urban sewage discharge and the treatment process used in the urban sewage treatment plant. However, as we know, the environment characteristics of rural areas are very different from that of urban. Rural areas have a lot of permeable land full of nature soils and plants whereas urban has a lot of building and impervious area. These different environmental characteristics in rural areas require unique treatment system and cannot simply copy the practice of urban. The system should be focused on the ecological treatment process of using soil and plants for sewage treatment. Unfortunately, the application of ecological treatment processes in rural area nowadays is still very limited. Thus, research on ecological treatment processes for rural areas needs to be strengthened; 2) The rural decentralized treatment facilities are usually installed at a site far away from the place where sewage is

Content from this work may be used under the terms of the Creative Commons Attribution 3.0 licence. Any further distribution of this work must maintain attribution to the author(s) and the title of the work, journal citation and DOI.
Published under licence by IOP Publishing Ltd

produced, and close to receiving water body like rivers and lakes. The sewage treated by facilities is directly discharged into the receiving water body. However, In the United States, the treatment facilities are constructed at the source site of sewage. As a result, the treated sewage by facility is infiltrated into the soil through a dispersal system buried under the surface instead of discharging into surface water directly. As to water body, the nitrogen, phosphorus and other substances in the sewage are the source of pollution to cause eutrophication whereas to soil they are nutrient sources for plant growth.

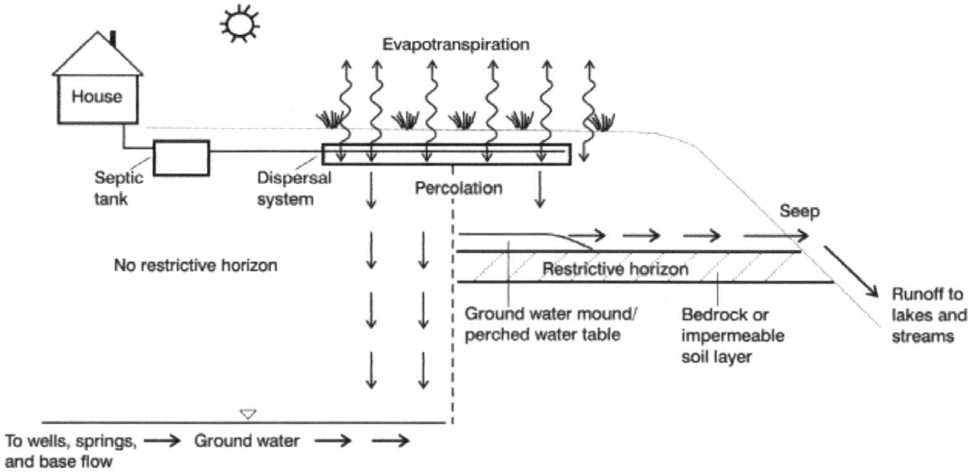

Figure 1. The decentralized treatment system of rural sewage in the United States

Therefore, we need to consider how to solve the above problems. Although many families in rural areas in China have built septic tanks, the water quality of household septic tanks built in rural China is usually not very stable and does not meet the water quality requirement for infiltrating into soil. It is difficult for us to copy the practice of the United States.Yan et al. have proposed a method to improve outflow quality by rebuilding septic tank and combining the use of soil percolation[3].However, this method is not easy to be widely applied in practice in rural areas. The reason is that reconstruction of septic tanks of various households is not only time-consuming and laborious, but also difficult to implement in Chinese rural areas.Therefore, how to learn from the practices in the United States and use soils and plants to turn pollution sources into nutrient sources is worthy of further study.

On the other hand, in the late 1990s, a new type of urban rainwater management and utilization method was developed in the United States, that is Low Impact Development, namely LID. Various kinds of LID technologies such as bioretention, grass ditch, etc. are employed to reduce storm runoff volume and rainwater non-point source pollution through storage, infiltration, filtration, and evaporation [4]. LID is a decentralized, source governance strategy, and this kind of governance strategy is considered to be suitable for solving both rainwater problem and rural sewage problem. There are many large and small soil ditches in rural areas of China, sewage produced by each family flows along these ditches and drain to receiving water body eventually. However traditional rural ditch usually has very low permeability, causing sewage to easily flow out from ditch and little infiltrate into soil through ditch bed to get purified. To overcome this issue of existing ditch, we propose a new type of ditch based on Low Impact Development (LID) strategy by building several weirs along ditch, replacing the bed soil of about 30 cm of thickness in the traditional ditch with filter materials, and covering a layer of coarse stone or pebbles on the top to avoid loss of filter materials.This new type of ditch is so-called new LID-type ditch having the functions of storage, infiltration, purification and drainage. The core component of new LID-type ditch is the filter material which needs to consider its purification performance, permeability and economy. Thus, this study carried out experimental comparisons on the purification ability and permeability of various filter materials produced in our laboratory, and briefly compared the cost of the filter materials. The study results provide a new way to treat rural sewage at the source site which is effective, ecological and economical.

2. Materials and method

2.1. Materials

It is better to choose filter materials which are not only cheap and easy to obtain but also having good purification ability. By consulting the related literature, four kinds of basic filter materials, Xiamen natural red loam, sand, gravel and zeolite are selected as the test materials.

2.1.1 Xiamen local natural red loam soil: It is the typical soil type in Fujian Province which has good adsorption to phosphorus and may reach a adsorption amount of 1.61mg/g [5]. Red loam soil is iron and aluminum-rich, the amorphous iron and aluminum content in red soil effectively promote absorbing phosphorus by chelation[6]. However, its permeability coefficient is only about 0.098mm/min [7], which is much less than the recommended infiltration rate (>1.143mm/min) based on experience by Prince George's County [8], thus its permeability needs to be improved artificially by adding some sands, gravel and etc.

2.1.2 Sand: It has a good adsorption effect on COD. The adsorption of COD by sand is mainly physical adsorption.Its large specific surface area can promote the contact adhesion of COD[9]. Xiamen local sand with particle size of 0.2-0.5mm is selected as test material.

2.1.3 Gravel: It has a good adsorption effect on TN, its adsorption is the main reason for the removal of TN from water [10].The gravel with particle size of 4-8mm which is the products of Miaoyuan Water Treatment Material Co., Ltd in Gongyi City of China is selected as test material.

2.1.4 Zeolite: It is a hydrous silicate mineral with a cluster-like structure. Cavities and pores with a certain pore size in the zeolite framework create its adsorption and ion exchange properties [11]. Zeolite has a good effect on the removal of ammonia nitrogen due to its chemical adsorption and ion exchange on ammonia nitrogen [12]. The gravel with particle size of 4-8mm, purchased from Miaoyuan Water Treatment Material Co., Ltd in Gongyi City of China, is selected as test material

The above 4 kinds of basic materials are used to generate 3 types of mixed filter materials by mixing these basic materials together with different mass ratios as shown in table 1. From type 1 to 3, the total mass ratios of red loam and sand decrease from 80% to 40% and that of gravel and zeolite increase from 20% to 60%.

Table 1. Mass ratios of various basic materials for 3 types of mixed filter materials.

type	Red loam(%)	Sand(%) (0.2-0.5mm)	Gravel(%) (4-8mm)	Zeolite(%) (4-8mm)
1	50	30	10	10
2	30	30	20	20
3	20	20	30	30

2.2. Experimental device

In order to compare the purification effects of 3 types of mixed filter materials, an experimental device as shown in Figure 2, is employed to fill 3 types of filter materials separately. Diameter of cylinder included in the device is 20cm and the height is 60cm. The device is mainly composed of 4 parts which are material filling cylinder, seepage board, overflow hole, water outlet. Seepage board is enlarged to be shown in Figure 3.

2.3. Indicators and determination methods

2.3.1. Water quality Indicators and determination methods: According to the monitoring indicators of rural sewage commonly used in Fujian Province of China, five water quality indicators are selected as monitoring indicators for our experiment which are total phosphorus (TP), ammonia nitrogen (NH_3-N), total nitrogen (TN), chemical oxygen demand (COD) and five-day biological oxygen demand (BOD_5).

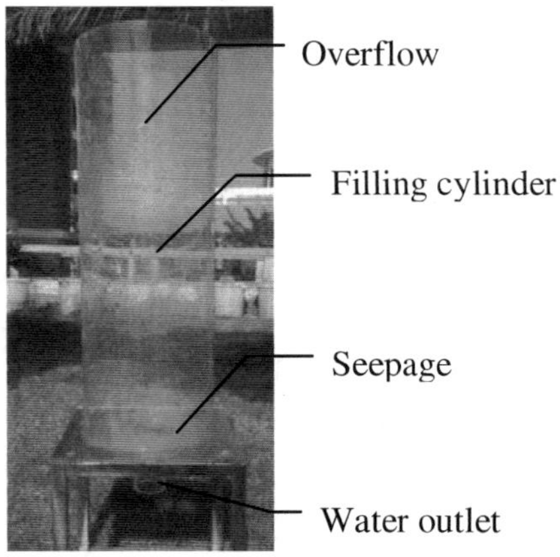

Figure 2. Experimental device.

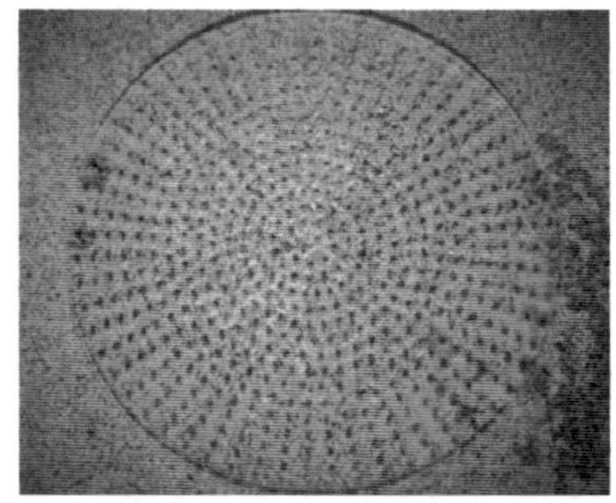

Figure 3. Seepage board

The determination methods and instruments for these five indicators are shown in table 2 and table 3 respectively.

Table 2. Water quality indcators and its corresponding determination methods

Water quality indicators	Determination methods
TP	High-pressure digestion of potassium persulfate, Stannous chloride spectrophotometry
TN	Alkaline potassium persulfate digestion, ultraviolet spectrophotometry
NH_3-N	Nessler reagent spectrophotometry
COD	Rapid digestion spectrophotometry
BOD_5	Dilution and inoculation

Table 3. Water quality indicators and corresponding determination instruments

Water quality indicators	Determination instruments
TP, TN	Palintest Digital Digester, Palintest-7100 Photometer
NH_3-N	Palintest-7100 Photometer
COD	LianHua 5B-I(V8) Intelligent Multi-Parameter Digestion Reactor, Lianhua 5B-3B(V8) Multi-Parameter Water Analyzer
BOD_5	Boxun Model SPX-250B-Z Biochemical Incubator, Hach BODTrak Water Quality Analyzer

2.3.2. Permeability Indicator and determination methods: Saturated permeability coefficient of K_s is used as an indicator to evaluate the permeability of the mixed filter materials. K_s is measured based on fixed head measurement method by using Model IK-4012 Saturated Soil Infiltration Instrument made in Japan.

IOP Conf. Series: Earth and Environmental Science **356** (2019) 012018 doi:10.1088/1755-1315/356/1/012018

2.4. Experimental method

1) Produce three types of mixed filter materials by evenly mixing 4 kinds of basic materials which are red loam, sand, gravel and zeolite based on the corresponding mass ratios shown in table 1. The mass of all these three types of mixed filter materials are controlled to be equal to 12.3kg.
2) Fill each type of mixed filter material to the experimental device shown in Figure 2 one by one. The heights of mixed filter materials filled in cylinder are 30cm, 30.5cm, 31cm for type 1, type 2 and type 3 respectively.
3) Collect rural domestic sewage from main drainage ditch of sewage in Yincun village of Houxi town nearby our university, and slowly pour the same volumes of sewage of 10L into three experimental pipes respectively.
4) Collecte the outflow water samples from the outlet shown in Figure 2.
5) Determinate the five water quality indicators of inflow and outflow, TP, TN, NH_3-N, COD and BOD_5, based on the method stated in 2.3.1.
6) Measure the saturated permeability coefficient of K_s based on the method stated in 2.3.2.
7) Analyse the experimental results.

3. Results and analysis

Purification experimental results for the mixed filter material of type 1, type 2 and type 3 are shown in table 4, table 5 and table 6 respectively. Comparison of the removal rate of each water quality indicator for 3 types of mixed filter materials are shown in Figure 4. The average removal rates for five water quality indicators of TP, TN, NH_3-N, COD and BOD_5 and saturated permeability coefficients of K_s for three types of mixed filter materials are shown in table 7.

Table 4. Purification experimental result for the mixed filter material of type 1 (unit: mg/l)

Water quality indicators / Water samples	TP	TN	NH_3-N	COD	BOD_5
Inflow of type 1	1.98	33	7.4	179.1	97.6
Outflow of type 1	0.37	23	5	87.3	58.4
Removal rates (%)	81.31	30.30	32.43	51.26	40.16

Table 5. Purification experimental result for the mixed filter material of type 2 (unit: mg/l)

Water quality indicators / Water samples	TP	TN	NH_3-N	COD	BOD_5
Inflow of type 2	2.95	36	7.8	21	10.9
Outflow of type 2	1.31	26	4.9	12	6.2
Removal rates (%)	55.59	27.78	37.18	42.86	43.12

Table 6. Purification experimental result for the mixed filter material of type 3 (unit: mg/l)

Water quality indicators / Water samples	TP	TN	NH_3-N	COD	BOD_5
Inflow of type 3	1.64	86	12.5	34.61	20.8
Outflow of type 3	0.61	45	6.8	27.09	13.4
Removal rates (%)	62.80	47.67	45.60	21.73	35.58

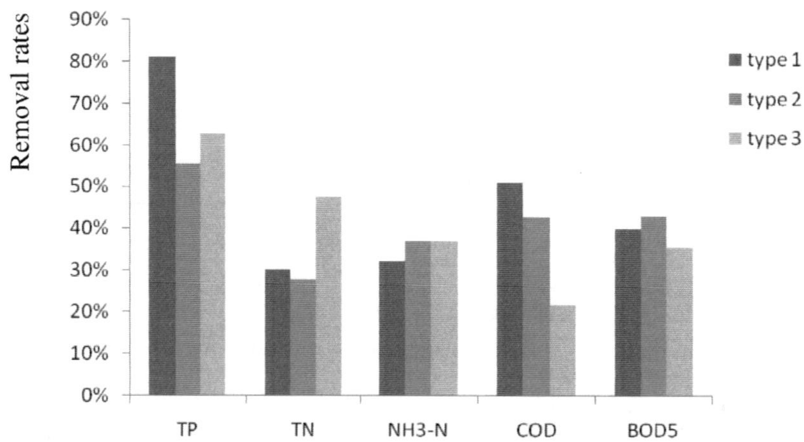

Figure 4. Comparions of removal rates for 3 types of mixed filter materials

Table 7. Average removal rates and K_s for three types of mixed filter materials

Type of mixed filter material	Average removal rate	Saturated permeability coefficient K_s Unit: m/s
Type 1	47.09%	$9.131*10^{-5}$
Type 2	41.31%	$2.889*10^{-4}$
Type 3	40.99%	$2.062*10^{-3}$

From table 4-7 and Figure 4, it can be seen that:

1) Type 1 has obvious advantages in removing TP and COD compared to type 2 and type 3. The reason is attributed to that the mass ratios of soil and sand for type 1 are the highest among three types of mixed filter materials while soil and sand have good purifications for TP and COD respectively [5,6,9].
2) Type 3 has obvious advantages in removing TN compared to type 1 and type 2. The reason is due to that the mass ratios of gravel and zeolite are the highest in three types of mixed filter materials while gravel and zeolite have good purifications for TN [10-11].
3) Type 2 has no obvious removing advantage for a certain water quality indicator. Since the sum of mass ratios of red loam and sand are at an average level among three types of mixed filter materials, and so do the sum of mass ratios of gravel and zeolite, thus, no outstanding performance can be seen for type 2.
4) The comprehensive purification capacity of type 1 is strongest with average removal rate up to 47.09% , while type 2 and type 3 have the same removal rate of approximately 41%.

Besides, form table 1, it can be known that the costs of three types of mixed filter materials follows this order in values, type 1< type 2 < type 3, since soil and sand are cheaper than gravel and zeolite, andfrom type 1 to 3the total mass ratios of red loam and sand decrease from 80% to 40% and that of gravel and zeolite increase from 20% to 60%.

Table 8 shows the discharge limits of the related five water quality indicators regulated in "Discharge standard of pollutants for municipal wastewater treatment plant (GB 18918-2002)" and "Farmland irrigation water quality standard (GB5084-2005)" in China, where the value of NH_3-N outside the parentheses is the limit when the water temperature is higher than 12°C, and the value in parentheses is the limit when the water temperature is lower than 12°C. Our experiments are conducted in winter and the temperature of water is lower than 12°C, thus the limits in parentheses should be applied in analysis.

Comparing the water quality conditions of outflow shown in table 4-6 to the standards shown in table 8, the following conclusions can be obtained.

1) Water quality of outflow for all of three types of mixed filter materials meets the requirement of both paddy and dry fields regulated by "Farmland Irrigation Water Quality Standard (GB5084-2005)".
2) Except the indicator of BOD_5, water quality of outflow for all of three types of mixed filter materials meets the requirement of level 2 discharge standard regulated by "Discharge standard of pollutants for municipal wastewater treatment plant.

Table 8. Dischareg limits of 5 related water quality indicators in two kinds of standards

Standards / Indicators	Discharge standard of pollutants for municipal wastewater treatment plant (GB 18918-2002) Unit: mg/L				Farmland irrigation water quality standard (GB5084-2005) Unit: mg/L	
	Level 1A	Level 1B	Level 2	Level 3	Paddy field	Dry field
TP	0.5	1	3	5	-	-
TN	15	20	-	-	-	-
NH3-N	5(8)	8(15)	25(30)	-	-	-
COD	50	60	100	120	150	200
BOD5	10	20	30	60	60	100

It can be also seen from table 7 that saturated permeability coefficients K_s of three types of mixed filter materials follows this order in values, type 1< type 2 < type 3. Take a LID-type ditch with a width of 1m and a length of 100m as an example, type 1 can treat 98.7m^3 of sewage under a condition of 3-hour infiltrating time per day, while type 2 and type 3 can treat 312.0m^3 and 2226.5m^3 of sewage under the same condition respectively. In rural areas, sewage is not discharged to ditch all the time. Usually, it is only produced by villagers within one time period such as 1 hour of morning, noon and evening. Therefore, 3 hours of full sewage supply per day is adopted to calculate infiltration in this example. However, the population of most of villages in China is less than 2000 people, the required design sewage treatment capacity of facility is usually less than 250m^3/d. It means, as long as a village's ditch has a total discharge length of more than 260m, the required design sewage treatment capacity can be met after being converted into a new LID-type ditch with the mixed filter material of type 1. Since the total length of village ditch is usually larger than 260m, therefore not only type 2 and type 3 but also type 1 are all applicable to the practices of rural sewage treatment.

4. Conclusions

A new LID-type ditch has been proposed in this study to treat rural sewage by replacing the bed soil in rural ditch with the mixed filter material for a thickness of about 30cm and setting several weirs along the ditch. Three types of mixed filter material have been generated by evenly mixing four basic common materials, red loam, sand, gravel and zeolite based on three different sets of material mass ratios. Their purification ability for TP, TN, NH_3-N, COD, BOD_5 and their saturated permeability coefficients have been analysed. The following conclusions can be drawn based on the study:

1) For purification ability, type 1 is better than type 2 and type 3 in general. The average removal rate for TP, TN, NH_3-N,COD, BOD_5 of type 1 achieves to 47.09% , while that of type 2 and type 3 have almost the same removal rate of approximately 41%. However, in particular, each type of mixed filter material has its own performance. Type 1 has obvious advantages in removing TP and COD, its removal rates of TP and COD achieve to 81.31%. Type 3 has an obvious advantage in removing TN with removal rate of 47.67%. While type 2 has no obvious removing advantage for a certain water quality indicators.
2) For permeability, type1<type2<type3. Although the saturated permeability coefficient of type 1 is smallest with a value of K_s=9.131*10^{-5}m/s, its permeability still can meet the required design

sewage treatment for most of Chinese village with a population of less than 2000. That means all three kinds of mixed filter materials are usually applicable in Chinese rural areas.

3) For economical performance, the total material costs of type 1< type 2 < type 3, i.e., type 1 is cheapest and type 3 is the most expensive.
4) In summary, type 1 of mixed filter materialwith 50% of red loam, 30% of sand, 10% of gravel, and 10% of zeolite is the best one for the application of LID-type ditches in rural areas based on the comprehensive comparison of purification, permeability and costs.

5. References

[1] United States Environmental Protection Agency (EPA) 2002.02 *USEPAOnsite Wastewater Treatment Systems Manual (EPA/625/R-00/008)* 1-7
[2] Song G J, Sheng YH 2006 *J. Environment and Sustainable Development***1** 20-23 (in Chinese)
[3] Yan Y N, Zhang L Y and Xi B D 2011 *J. China Water & Wastewater***27(10)** 69-72 (in Chinese)
[4] United States Environmental Protection Agency2000*R.Low Impact Development (LID): A Literature ReviewEPA-841-B-00-005*, Washington DC
[5] Huang Z Z, Wu X F and Zhao F 2009 *J.China Water & Wastewater***25(21)** 91-94 (in Chinese)
[6] Hunt W F, Davis A P and Traver R G 2012*J. Environ. Eng.138* (6), 698-707.
[7] Zhu M L, Liao J, Chen G Y, Wang J P and She N 2009*J. Water Resources Protection***29(3)** 25-28(in Chinese)
[8] County P G1999 Low-impact development design strategies: An integrated design approach. *Prince George's County, MD Department of Environmental Resources.*
[9] Wang J L, Qin Q C, Zhang Y Y, Zhang Y J, Feng C M, Xu P and Li J Q 2016 *J. Science Technology and Engineering***9** 86-92 (in Chinese)
[10] Wu H 2006. *D. Experimental study on removal of nitrogen from water by gravel riverbed* (Nanjing: Hohai University) (in Chinese)
[11] Ortega E, Cheeseman C and Knight J 2000 *Cement and Concrete Research***30** 1641-1646
[12] Zhang X, Wu W Z, Wen D H, Li W Q and Tang X Y 2003 *Environmental Chemistry* **2**166-171 (in Chinese)

6. Acknowledgments

This work was financially supported by "Construction Science and Technology Project of the Xiamen Construction and Administration Bureau (XJK2018-1-2)" and "Fujian Water Conservancy Science and Technology Project of Fujian Provincial Department of Water Resources (MSK201713)"

Control and cleaning of membrane biofouling by biogenic silver nanoparticles

Manying Zhang[1]*, Cheng Cheng[1] and Rongrong Xie[1]

1.School of Chemical and Environmental Engineering, Jiangsu University of Technology, Changzhou 213001, China
E-mail: myzhang@jsut.edu.cn

Abstract. Membrane biofouling as the main obstacle strongly hinders the wider application of membrane technology. The control of biofouling development and the fouled membrane cleaning still are the big challenge in membrane industry. The present study investigated the control of initial bacterial attachment, biofilm development and finally the membrane cleaning by biogenic silver nanoparticles (Bio-Ag^0-6) using commercial microfiltration membranes as the substrate. The results revealed that the Bio-Ag^0-6 could inhibit the development of biofilm by decreasing the attachment of cells and the production of extracellular polymeric substances (EPS). Significant improvements of flux recovery ratio (FRR), the distribution of fouling resistances and the decrease of total fouling resistance (R_f) were achieved when introducing Bio-Ag^0-6 to dead-end microfiltration systems. The membrane cleaning experiment verified the enhanced biofilm detachment from membrane surface by Bio-Ag^0-6. It shows that Bio-Ag^0-6 would be a good alternative to relieve membrane biofouling.

1. Introduction

Membrane biofouling is a major challenging issue of membrane application in water industry[1,2]. Biofouling, which is considered more problematic than organic fouling forms through the initial deposition of bacterial cells on the substrate, accumulation of organic macromolecules in EPS, colony development and finally maturation of biofilm [3]. Biofouling is difficult to eliminate because microbes could fully utilize the nutrient in the feed to multiply rapidly. As the products of microbial aggregates, EPS could bridge cells together [4]. EPS played an important part in biofilm formation,and was considered to be the main reason of membrane biofouling [5]. Besides, it is considered that the initial irreversible attachment, the activity and proliferation of microorganisms all have great influence on the process of biofouling [6,7]. Biofouling results in the flux decline and increased operation and maintenance costs [8].

Common treatments are used to prevent the development of biofouling. Disinfectants such as ozone and chlorine or UV irradiation as pretreatment, reducing nutrients in the permeat or utilizing antifouling membrane materials to control the formation of biofilm. However, the continuous application of oxidative disinfectants may form disinfection byproducts and are detrimental to membrane[9].

Besides promoted biofilm detachment is another method to alleviate membrane fouling[10]. Many physical-chemical ways such as sonication, back-washing and chemical cleaning have been used to remove biofilm. However, microorganisms may develop resistance to the traditional processes and the mature biofilm is often hard to remove totally [11,12]. Moreover, conventional approaches to deal

Content from this work may be used under the terms of the Creative Commons Attribution 3.0 licence. Any further distribution of this work must maintain attribution to the author(s) and the title of the work, journal citation and DOI.
Published under licence by IOP Publishing Ltd

with membrane biofouling may be costly and time-consuming. Hence, the research of effective new methods to relieve biofouling has attracted tremendous interest.
Silver nanoparticles (AgNPs) as a broad spectrum of anti-microbial agent have been widely used in healthcare products, water purification and membrane filtration systems. Silver nanoparticles have been incorporated into nanofiltration (NF), reverse osmosis (RO) and ultrafiltration (UF) membranes [13-16] and other materials. The fabrication of silver nanocomposite membranes significantly improved the antifouling and filtration performances.
The chemical reduction methods are the most commonly used method to synthesize silver nanoparticles. However, the chemical silver nanoparticles often have poor particle stability and tend to aggregated easily [17].
Recently, biosynthesis of silver nanoparticles has attracted much attention. As reported before, we fabricated small sized (≈6 nm) novel biogenic silver nanoparticles (Bio-Ag^0-6). The results demonstrated that the small sized and highly dispersed Bio-Ag^0-6 were efficient to control bacterial and biofilm growth[18].
In this study, the small sized and highly dispersed Bio-Ag^0-6 nanoparticles were firstly introduced into the commercial microfiltration membrane system which was quite different from our previous reports. This study further survey the effect of Bio-Ag^0-6 on the microbial attachment and the biofilm formation on commercial membrane surfaces. Changes of the filtration flux and the distribution of membrane fouling resistances were both evaluated by dead-end filtration test. Furthermore, we studied the influences of biogenic silver on the detachment of established biofilms from membrane surfaces and compared the membrane cleaning performance with traditional chemical agents. Thus, it is expected that the small sized Bio-Ag^0-6 will be an efficient alternative to alleviate biofouling in commercial membrane systems.

2. Materials and methods

2.1. Model bacterial strain and media
Brevundimonas diminuta (BD) was grown in LB medium at 37°C and180rpm. The strain was chosen as the model bacterial because it is often found in biofilms and has been widely used to assess biofouling in membrane systems[19-21].

2.2. Fabrication and characterizeation of Bio-Ag^0-6
The detailed fabrication and characterization of Bio-Ag^0-6 can be found in our previous study [21].

2.3. Bacteria attachment assay
To investigate the influence of Bio-Ag^0-6 in controlling biofilm formation, the bacteria attachment was conducted in microtiter plates (Nunc, Demark) [22,23]. Mixed cellulose ester (MEC) membranes (0.1μm, VCWP02500 ; Millipore, USA) were the substrates for biofilm growth. BD was grown overnight and then diluted to get the working solution. The final silver concentrations were 0, 5, 10, 15mg/L. Finally the plate was cultivated at 37 ℃ for 72 h.
The membranes were fully immersed into phosphate buffer saline (PBS) solution to remove loosely adherent cells and then sonicated for 5min. The number of attached living bacteria on membrane samples was confirmed by gradient dilution and spread plate methods.

2.4. The analysis of EPS
EPS is mainly composed of proteins (PN) and polysaccharides (PS). After the 72-h-aged biofilms on MEC membranes were performed as described in the section headed 'Biofouling formation assay' above, EPS was extracted using methods similar to those reported before [24-25]. PN were assayed by BCA Protein Assay Kits (23227; Thermo, USA). PS were analyzed by anthrone-sulfuric acid method. BSA and glucose were used as the protein and polysaccharide standards.

2.5. SEM characterization

The morphologies of biofilms distributed on MEC membranes under treatment of Bio-Ag^0-6 were observed by scanning electron microscopy (SEM). The 72-h-aged biofilms were washed with PBS and treated with glutaraldehyde solution (2.5%) at 4 ℃ for 6 h. Before SEM observation, the samples were steply dehydrated and dried in an oven at 60 ℃ over 24 h [26].

2.6. Staining and CLSM imaging

The influence of Bio-Ag^0-6 pretreatment on the viability of attached cells and the PS distributions in biofouling layers were characterized by CLSM (LSM710, Germany). Propidium iodide (PI) and nucleic acid stains SYTO 9 (Ex=485nm, Em= 630/530 nm) was used to stained the biofilm [27]. PS was binded with Alexa Fluor conjugated Con A (Ex=590 nm, Em=617 nm) [28]. The membranes were stained in the dark for 30 min. Finally the membranes were washed with PBS and observed by CLSM.

2.7. Dead-end microfiltration test

The dead-end microfiltration test was conducted using filtration cell (Amicon 8010; Millipore) at 0.05MPa. The flux (J) was obtained with Equation (1):

$$J = \frac{V}{A\Delta t} \quad (1)$$

Where V (L) is the volume of permeated water, A (m^2) is the effective area and Δt (h) is the filtration time.

Then the biofouling experiment was performed. Following the 72 h biofouling, membrane sample was cultivated on LB agar to facilitate bacteria reproduction. The water flux (Jp) was measured again. In order to compare the antifouling performance of the membrane, the relative flux reduction (RFR) was calculated as [28,29]:

$$RFR = \left(\frac{J_{w1} - J_p}{J_{w1}} \right) \times 100\% \quad (2)$$

Furthermore, the fouling resistance (R_f) and the intrinsic membrane resistance (R_m) were determined by the resistance-in-series model [21,29]:

$$J = \frac{\Delta P}{\mu R_t} \quad (3)$$

where J is the water flux (m^3 m^{-2} s^{-1}); μ is the solution viscosity (Pa · s) and ΔP is the filtration pressure (Pa).

$$R_t = R_m + R_f \quad (4)$$

The R_m can be estimated from Equation (3) by measuring the flux of clean membrane, and R_t is obtainted from the fouled membrane water flux, thus R_f can be obtained.

2.8. Membrane cleaning

In order to compare the cleaning effect of different agents, the bacterial suspension filtration experiment was conducted by dead-end filtration.. After the initial pure water flux (J_{w1}) was obtained, 2 mL bacterial suspension was filtered without stirring to facilitate the deposition and adhesion of cells on membrane. Following filtration of BD suspension, each sample was placed on LB agar to enhance the bacteria multiplying. The water flux (J_p) after 24 h biofouling was determined again. Finally, the

samples were cleaned with Bio-Ag^0-6 (30 mg/L), sodium hypochlorite (NaClO, 200 mg/L) and PBS (0.1 M, pH 7.2) by shaking for 30 min, respectively. Then the flux (J_{w2}) was tested again.
The flux recovery ratio (FRR) can be got using Equation (5):

$$FRR = \left(\frac{J_{w2}}{J_{w1}}\right) \times 100\% \quad (5)$$

Moreover, the fouling resistance parameters were analyzed to study the fouling process [24]. The irreversible (Rir), reversible (Rr) and total fouling ratio (R_t) were calculated as:

$$R_t = \left(1 - \frac{J_p}{J_{w1}}\right) \times 100 \quad (6)$$

$$R_r = \left(\frac{J_{w2} - J_p}{J_{w1}}\right) \times 100 \quad (7)$$

$$R_{ir} = \left(\frac{J_{w1} - J_{w2}}{J_{w1}}\right) \times 100 \quad (8)$$

3. Results and discussion

3.1. Effect of Bio-Ag^0-6 on BD attachment

The cell numbers during 72 h incubation are plotted as a function of different Bio-Ag^0-6 concentrations in Figure1. After 72 h of cultivation, the strain BD displayed excellent biofilm formation property. As shown in Figure 1, without Bio-Ag^0-6 pretreatment, BD showed significant growth after 72 h incubation with the highest log CFU of about 7.0. Bio-Ag^0-6 of 5 mg/L slightly decreased the number of live bacterial. The more Bio-Ag^0-6 content increased, the more obvious the decrease effect became. As for 10 mg/L and 15 mg/L of silver, there were over four and five orders of magnitude decrease for BD, respectively. The figure showed that the Bio-Ag^0-6 significantly affected the multiplying and attachment of BD on MEC membranes, which was probably due to the excellent antibacterial and antifouling effect of Bio-Ag^0-6 as reported before [30].

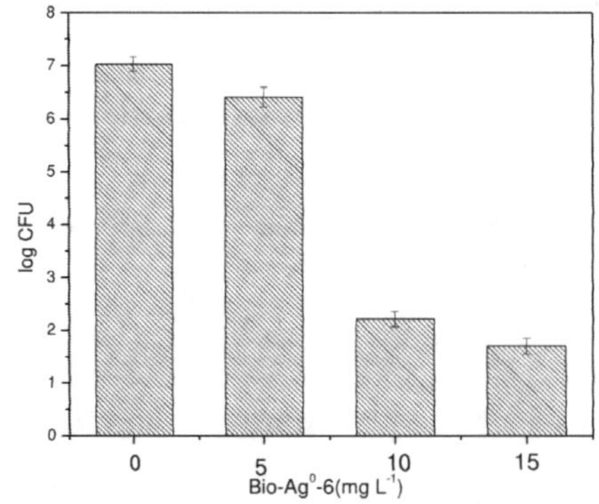

Figure 1. The cell numbers of BD biofilms on MEC membranes after 72 h of cultivation.

3.2. EPS analysis

EPS is considered as the source of organic fouling and play an important part in membrane fouling process. EPS are composed of polymers produced by bacteria, such as humic substances, proteins and carbohydrates [22]. The concentrations of proteins (PN) and polysaccharides (PS), as the most important components in biofilms are shown in Table 1. As for the control sample (0 mg/L), the concentrations of PS and PN were 431.12μg/cm^2 and 250.07μg/cm^2, respectively. The PS contents were generally higher than PN which was consistent with reports before [22,31]. For all the samples, the productions of PS and PN were inversely proportional with the Bio-Ag^0-6 content. As for 15 mg/L of silver, the concentrations of PS and PN dropped dramatically to 57.75μg/cm^2 and 34.72μg/cm^2, respectively. The results indicated that lower concentration of EPS may be an useful method to relieve membrane biofouling [31].

Table 1. The concentrations of polymers in EPS in 72h aged biofilms.

Bio-Ag^0-6 (mg/L)	polysaccharides concentration(μg/cm^2)	proteins concentration (μg/cm^2)
0	431.12±6.52	250.07±3.22
5	284.10±4.38	205.97±2.75
10	201.99±2.96	162.11±2.28
15	57.75±1.63	34.72±0.76

3.3. SEM and CLSM images of biofouling layer

To further monitor the biofilm formation under different concentration of Bio-Ag^0-6, the SEM characterization was carried out. As for the control sample in Figure 2, BD displayed strong biofilm formation ability. The membrane surface was fully covered by bacterial clusters and completely fouled. Quite different from the control sample, much fewer cells were observed on the membrane even when the Bio-Ag^0-6 concentration was only 10 mg/L. Most of the foulants including EPS had disappeared on the membrane surfaces.

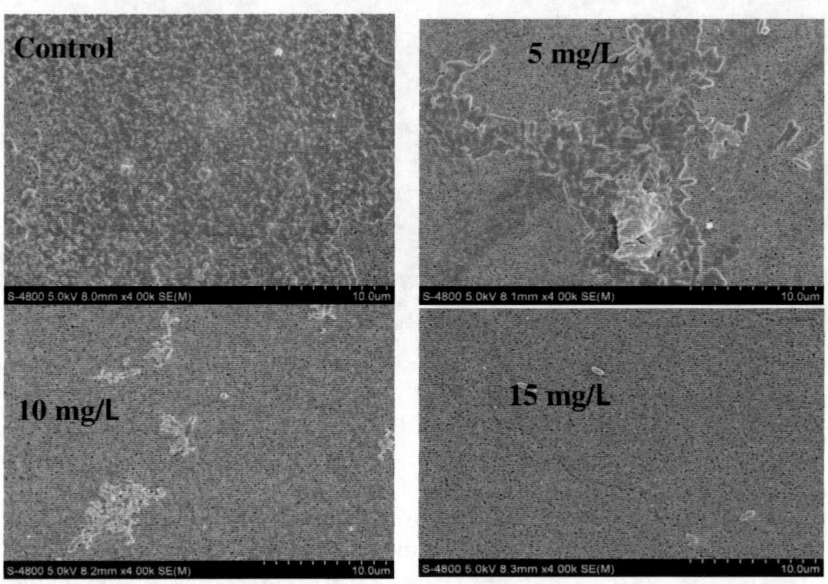

Figure 2. SEM images of fouling layers after 72h growth at different Bio-Ag^0-6 concentrations.

Multicolor fluorescence labeling and CLSM observation were used to characterize the composition and structure of biofouling layers. Figure 3 showed the fluorescent labeling graphics of control (without Bio-Ag^0-6) and biofouling samples grown in 10 mg/L of Bio-Ag^0-6 as an example. As shown

in Figure 3a1, numerous green clusters evenly dispersed on the MEC membrane surface, indicating that almost all the bacteria in the control biofilm were alive. As for the treated samples (Figure 3a2), the total coverage of cells decreased dramatically. Very few cells, most of which showed as small red dots were present, indicating that most cells were dead. Furthermore, the concentration and distribution of organic macromolecules such as polysaccharides (PS) in EPS became visible. A large number of polysaccharides clusters were observed (in Figure 3b1) in control biofouling layer, whereas a significant decrease in the production of polysaccharides (in Figure 3b2) was found on the membrane samples cultivated in 10 mg/L of Bio-Ag^0-6. This is in agreement with the SEM results in Figure 2.. This is also consistent with the above EPS analysis in Table 1.
EPS which is defined as a bio-glue to maintain the 3D structure of a biofilm is the main contributors to biofouling [28]. All the findings above implied that the pretreatment of Bio-Ag^0-6 could remarkably reduce the initial attachment of biofilm on the membrane surfaces and the production of EPS, thus may have a positive impact on the membrane biofouling control.

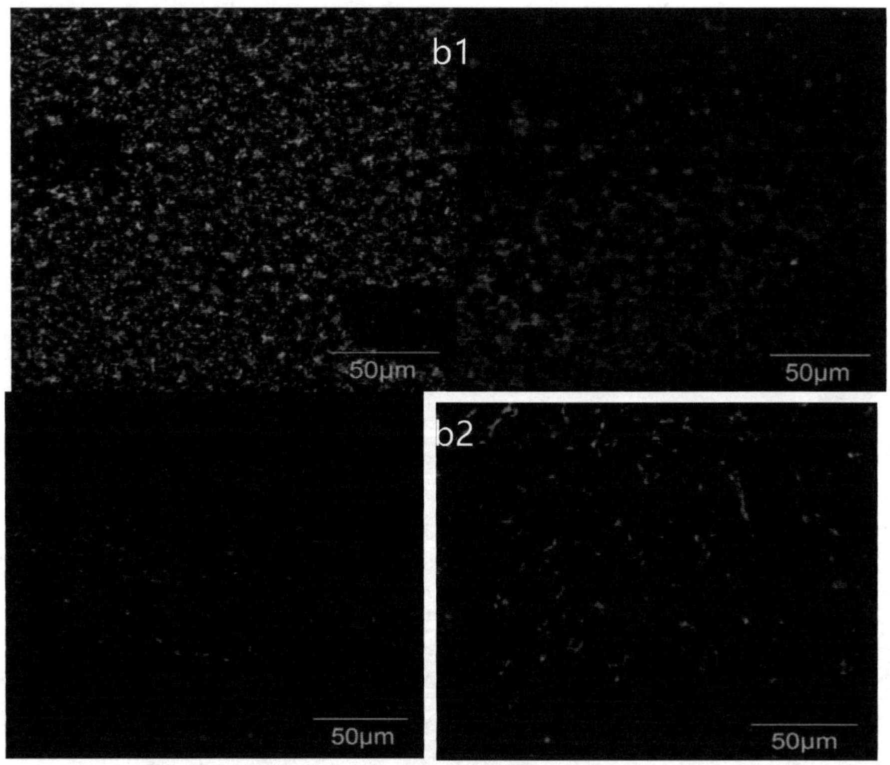

Figure 3. CLSM images for 72-h old biofouling layers of control (a1-b1) and samples cultivated in 10 mg/L of Bio-Ag^0-6 (a2-b2). (a) Adherent biomass (live cells in green; dead cells in red); (b) PS (polysaccharides in magenta).

3.4. Membrane flux
The membrane biofouling tests were conducted using dead-end filtration cell to compare the flux differences that caused by the application of Bio-Ag^0-6. As shown in Figure 4, the original flux (J_{w1}) of all samples were over 1000 L $m^{-2}h^{-1}$. After the bacteria attachment and 24 h cultivation, all the flux (J_p) decreased. The flux of MEC membrane without Bio-Ag^0-6 addition declined almost 73% compared with its original pure water flux. With the concentration increase of Bio-Ag^0-6, the flux decline was only 52%, 29% and 8% for 5 mg/L, 10 mg/L and 15 mg/L of Bio-Ag^0-6 addition, which was relatively lower than the control MEC sample. This may due to the bacteria multiplying and biofilm sedimentation on membrane surfaces.

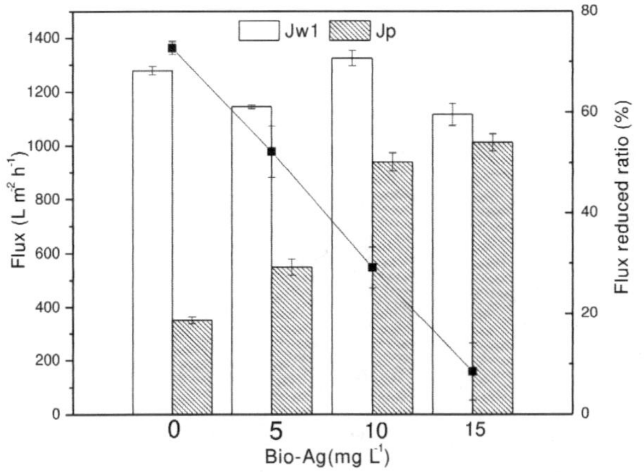

Figure 4. The water flux of membranes with Bio-Ag^0-6 pretreatment before and after biofouling (bars) also the flux decline of samples (points).

To further prove the superiority of the Bio-Ag^0-6 addition to the fouling control, the resistances after biofilm formation were calculated as described in Equation 3. As shown in Figure 5, all the MEC membranes have the similar intrinsic membrane resistance (R_m) of about 15 (10^{10} m^{-1}). The small errors were caused by the nomal inherent differences of all the MEC membranes. The values of R_f caused by the fouling increased a lot with the decrease of Bio-Ag^0-6 concentration, which may result from the biofouling and the membrane pore entrapped by bacteria [30]. Figure 5 also showed that the total filtration resistance (R_t), which is the sum of R_m and R_f was negatively correlated to the content of Bio-Ag^0-6. This is consistent with the change of R_f and the permeate flux in Figure 4.

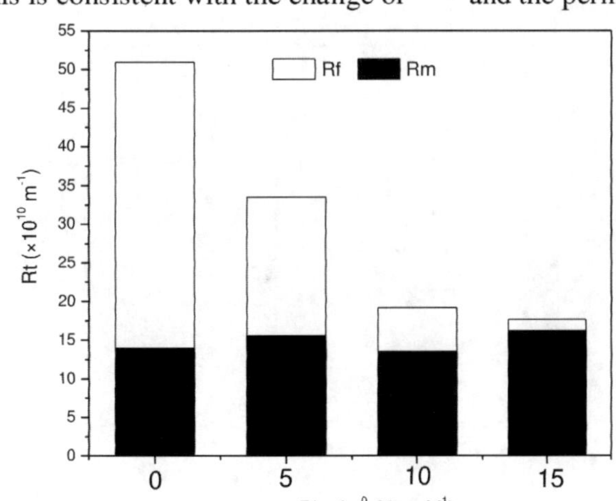

Figure 5. The distribution of intrinsic membrane resistance (R_m) and fouling resistance (R_f) in total resistance (R_t) of all the membranes with different Bio-Ag^0-6 dosages.

It has been known that the Bio-Ag^0-6 could damage the cells and showed great effectiveness in controlling the biofilm formation due to the release of biotoxic Ag^+ to the surrounding [29]. The results above suggest that the addition of small sized and highly dispersed Bio-Ag^0-6 may play a positive role for membrane biofouling control.

3.5. Membrane cleaning effect

To investigate the membrane cleaning effect using different cleaning agents, the dilute BD solution permeation process was conducted. Figure 6 shows the flux recovery ratio (FRR) after treated by different reagents. A more significant FRR was obtained after the fouled MEC membrane was soaked in Bio-Ag^0-6 solution (30 mg/L) than in PBS solution and NaClO (200 mg/L) . For example, the FRR for PBS treated membrane was only 12 %. Nevertheless, after soaking in NaClO solution, the FRR was increased to 47.7 % and further to 53.2 % after soaking in Bio-Ag^0-6 solution under the same condition. This also indicates that Bio-Ag^0-6 is more efficient to promote the detachment of biofilms from membrane surface, leading to the improved water flux.

To quantitatively assess the membrane cleaning effect of different agents, fouling resistance ratios were obtained as shown in Figure 6. Obviously, R_t was the sum of R_r and R_{ir}. Rr represents the flux decline caused by the gel layer formation and reversible foulant attachment, which could be wiped off by membrane cleaning. While Rir represents the flux decline induced by pore blocking and strong sedimentation of biofoulants in the pore or on the membrane surface [31]. Figure 6 showed that all the MEC membranes had the similar total fouling ratio after the biofouling was maximized. However, the proportion of R_r and R_{ir} in the total fouling ratio was quite different. When using PBS solution as the cleaning agent, the R_t of MEC membrane was made up of nearly 87.9% Rir and 10.4% R_r. As for 200 mg/L of NaClO solution, the Rir significantly reduced and the Rr increased. The 30 mg/L of Bio-Ag^0-6 treated MEC membrane has the highest reversible resistance ratio (Rr) of about 52% and the lowest irreversible resistance ratio (Rir) of about 46.8%. Larger Rr indicated that more adherent foulants could be removed more easily by Bio-Ag^0-6 washing. The reduction of Rir also directly confirmed that the membrane cleaning effect was elevated a lot by Bio-Ag^0-6 compared with the traditional cleaning agents.

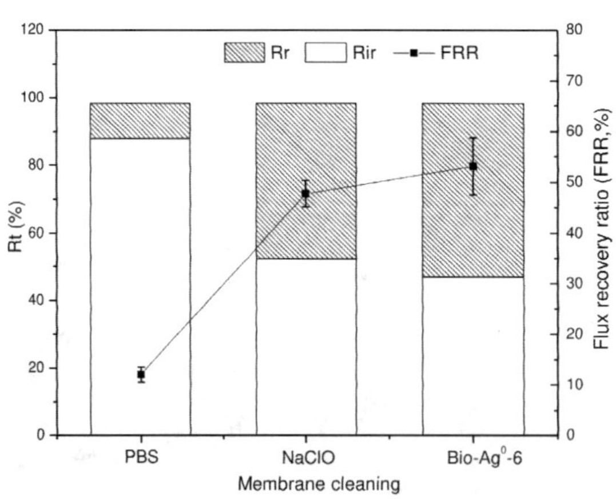

Figure 6. The flux recovery ratio (points) and the fouling resistance ratio of MEC membranes (bars) under different membrane cleaning methods.

4. Conclusions

In the present work, the small sized Bio-Ag^0-6 were introduced into commercial membrane systems. The biofouling formation assay and SEM results showed that the Bio-Ag^0-6 could affect the growth and the initial attachment of BD on MEC membranes. The EPS analysis and the multicolor fluorescence labeling in CLSM observation were consistent, revealing that Bio-Ag^0-6 showed excellent antibacterial activity, effectively reduced the EPS production and biofilm formation. The permeate flux changes and the fouling resistance calculation showed that the pretreatment of Bio-Ag^0-

6 could control the development of biofouling and decrease the membrane fouling resistance caused by fouling (R_f) in commercial membrane system. Moreover, it was much more efficient in facilitating the detachment of rigid biofilm from commercial membrane surface. From the results above, it is considered that the highly dispersed Bio-Ag^0-6 nanaoparticles could be used to relieve membrane biofouling, leading to the development of commercial membrane fouling control and effective membrane cleaning technology in membrane industry.

5.References

[1] A. Agarwal, W.J. Ng and Y. Liu 2012 *Biofouling* **29** 69-76.
[2] A. Nguyen, L. Zou and C. Priest 2014 *J. Membr. Sci* **454** 264-271.
[3] A. Ananth, G. Arthanareeswaran, A. F. Ismail, Y. S. Mok and T. Matsuura 2014 *Colloids Surf A Physicochem Eng Asp* **451** 151-160.
[4] C. Yu, J. Wu, A. E. Contreras and Q. Li 2012 *J. Membr. Sci* **423**, 487-494.
[5] C. Dong, G. He, H. Li, R. Zhao, Y. Han and Y. Deng 2012 *J. Membr. Sci* **387**, 40-47.
[6] D. Y. Koseoglu-Imer, B. Kose, M. Altinbas and I. Koyuncu 2013 *J. Membr. Sci* **428**, 620-628.
[7] G. E. Chen, H. H. Huang, Z. L. Xu, P. Y. Zhang, W. Z. Wu, L. Sun and Y. J. Liu 2016 *Membr. Water Treat* **7**, 539-553.
[8] H. C. Flemming 2002 *Appl. Microbiol. Biotechnol* **59** 629-635.
[9] H. C. Flemming, T. R. Neu and D. J. Wozniak 2007 J. Bacteriol **189** , 7945-7952.
[10] H. Xu and Y. Liu 2010 *Environ. Sci. Technol* **45** 595-601.
[11] H. Elcik and M. Cakmakci 2017 *Membr. Water Treat* **8** 499-515.
[12] H. B. Li, W.Y.Shi, Y.H.Su, H.X. Hou, Q.Y.Du, H.X.Zhang and X.H.Qin 2017 *Membr. Water Treat* **8** 225-244.
[13] J. Kim and B. Van der Bruggen 2010 *Environ. Pollut*, **158** 2335-2349.
[14] J. S. Teodósio, M. Simões, L. F. Melo and F. J. Mergulhão 2010 *Biofouling* **27** 1-11.
[15] J. Yin, Y. Yang, Z. Hu and B. Deng 2013 *J. Membr. Sci* **441** 73-82.
[16] J. H. Li, X. S. Shao, Q. Zhou, M. Z. Li and Q. Q. Zhang 2013 *Appl. Surf. Sci* **265** 663-670.
[17] M. Sile-Yuksel, B. Tas, D.Y. Koseoglu-Imer and I. Koyuncu 2014 *Desalination* **347** 120-130.
[18] M. Safarpour, V. Vatanpour and A. Khataee 2016 *Desalination* **393** 65-78.
[19] M. Zhang, K. Zhang, B. De Gusseme, W. Verstraete and R. Field 2014 *Biofouling* **30** 347-357.
[20] P. Chen, L. Cui and K. Zhang 2015 *J. Membr. Sci* **473** 36-44.
[21] P. T. P. Aryanti, M. Sianipar, M. Zunita and I. G. Wenten 2017 *Membr. Water Treat* **8** 463-481.
[22] S. Kwon, E. Moon, T.S. Kim, S. Hong and H.D. Park 2011 *Microbes Environ* **26**, 149-155.
[23] W. Xu and S. Chellam 2005 *Environ. Sci. Technol*, **39** 6470-6477.
[24] X. Li, R. Pang, J. Li, X. Sun, J. Shen, W. Han and L. Wang 2013 *Desalination* **324** 48-56.
[25] Y.S. Kang, Y.H. Kim, D.K. Lee, H.G. Cha and C.W. Kim 2007 *J. Phys. Chem. C* **111** 3629-3635.
[26] Y. Lequette, G. Boels, M. Clarisse and C. Faille 2010 *Biofouling* **26** 421-431.
[27] Y. Liu, E. Rosenfield, M. Hu and B. Mi 2013 *Water Res.* **47** 2949-2958.
[28] Y.Zhang, X. Yu, S. Gong, C. Ye, Z. Fan and H. Lin 2014 *Appl. Microbiol. Biotechnol* **98** 1309-1405.
[29] Y. Jafarzadeh and R. Yegani 2015 *Chem. Eng. Res. Des* **93** 684-695.
[30] Y. Zhang, L. Wang and Y. Xu 2015 *Desalination* **358** 84-93.
[31] Z. Wang and Z. Wu 2009 *Sep. Sci. Technol* **44** 3571-3596.

Acknowledgments

This work was supported by the National Natural Science Foundation of China (51508239); Postgraduate Education Reform Project of Jiangsu Province (JGLX18-165、JGLX18-166); Postgraduate Research & Practice Innovation Program of Jiangsu Province (SJCX18_0978) and Qing Lan Project.

IOP Conf. Series: Earth and Environmental Science **356** (2019) 012020 doi:10.1088/1755-1315/356/1/012020

Modeling Climate Change Impacts on Water Resources

X.W. Zhuang[1], Y.P. Li[2] and S. Nie[3] and G.H. Huang[4]

1 Institute for Civil Engineering, Qingdao Huanghai University, Qingdao 266427, China
2 State Key Laboratory of Water Environment Simulation, School of Environment, Beijing Normal University, Beijing 100875, China
3 Faculty of Applied Science and Engineering, University of Toronto, Toronto, ON M5S 1A4, Canada
4 Institute for Energy, Environment and Sustainable Communities, University of Regina, Regina, Saskatchewan S4S 0A2, Canada
Email: zhuangzi1989@126.com

Abstract. In this study, the effect of changes in daily temperature (minimum temperature and maximum temperature) during the baseline period on daily streamflow has been analyzed. The stepwise cluster analysis method is able to use a generated cluster tree to express the complex and nonlinear relationships between predictors and predictands. The results can be found that different characteristics of air temperature and precipitation exist during different seasons. Besides, the change of water resources can be indicated and is affected both by changes of climate factors and economic consequences. The most significant findings are that the simulated streamflow under various downscaling climate projections shows an increasing trend from future to recent periods.

1. Introduction

Climate change refers to changes in the mean and/or the variability of climate's properties that persists for an extended period, typically decades or longer, due to natural variability or as a result of human activity[1]. In terms of hydrology, changing of temperature and precipitation can pose direct influences on the quantity of evapotranspiration, both quality and quantity of the runoff component. Global climate models (GCMs) are the most powerful tools available to simulate evolving and future changes in the climate system, and can generate climate change scenarios to evaluate how hydrological systems may be affected by climates change. However, challenges are confronted that the coarse spatial resolution (hundreds of kilometres) of GCMs output, means that the output is not appropriate for direct use with local-scale impact model[2]. Consequently, it is important to understand the impacts of climate change on water resources, and to develop adaptive capacity to respond to these impacts.

2. Methodology

In this study, a stepwise statistical downscaling method is advanced for constructing local-scale climate change projections from large-scale GCMs[3, 4]. More detailed description of the stepwise cluster analysis method can be referred to Huang et al. (2006)[5]. The daily climate-streamflow hydrological model can be employed for streamflow simulation to assess impacts of climate change on the hydrology, using downscaled climate projections as inputs[6]. The simulated results during calibration and validation can be checked by using the Nash-Sutcliffe Efficiency (*NSE*)[6], and coefficient of correlation (R^2)[7].

Content from this work may be used under the terms of the Creative Commons Attribution 3.0 licence. Any further distribution of this work must maintain attribution to the author(s) and the title of the work, journal citation and DOI.
Published under licence by IOP Publishing Ltd

$$NSE = 1 - \frac{\sum_{i=1}^{n}(Q_i - C_i)^2}{\sum_{i=1}^{n}(Q_i - \bar{Q})^2} \quad (1a)$$

$$R^2 = \frac{[\sum(obs - \overline{obs})(pred - \overline{pred})]^2}{\sum(obs - \overline{obs})^2 \sum(pred - \overline{pred})^2} \quad (1b)$$

3. Result Analysis

The observations of three climate variables (i.e., maximum-temperature, minimum-temperature and precipitation) for the baseline (1981-2010) period are used as predictands in the study. Figure 1 shows the patterns of daily mean temperature, maximum temperature and minimum temperature during 1981-2010[8]. For instance, daily mean temperatures range from about -0.6 ℃ to 8.5 ℃ in the winter seasons; daily mean temperatures range from about 8.7 ℃ to 22.8 ℃ in the spring seasons; daily mean temperatures range from about 22.3 ℃ to 25.9 ℃ in the summer seasons; and daily mean temperatures range from about 0.5 ℃ to 23.0 ℃ in the autumn seasons. In comparison, the daily maximum temperature ranges from 2.7 ℃ to 28.8 ℃; the daily minimum temperature ranges from -3.2 ℃ to 23.8 ℃. Figure 2 presents the monthly air temperature, including monthly mean temperature, maximum temperature and minimum temperature during 1981-2010[9].

Figure 3 shows patterns of daily precipitation during 1981-2010. In detail, the maximum amount of precipitation is 11.2 mm which occurred in summer season. The minimum amount of precipitation is 0 mm which occurred in spring, autumn and winter seasons. The characteristics of monthly precipitation during 1981-2010 can be indicated in Figure 4. In detail, the monthly mean precipitation in August was 151.9 mm. From 1981 to 2010, the maximum amount of precipitation in August was 482.2 mm which occurred in 2007. In comparison, the minimum amount of precipitation in August was 29.2 mm which occurred in 2002. Besides, the monthly mean precipitation, maximum precipitation and minimum precipitation in January would be 10.8 mm, 40.8 mm and 0 mm, respectively. Furthermore, winter precipitation will decrease and temperature will increase leading to earlier melt snow contributing to soil moisture saturation[10]. Generally, an increase trend of the air temperature can be indicated, and the different patterns of precipitation would result a significant changes in water resources.

Figure 5 presents the water shortage amount under case 1 ($q = 0.01$), when the streamflow is at different levels. In detail, the water shortage amounts during period 1 ($t = 1$, 2016-2020) would be [843.8, 1816.1] (low), [564.0, 1451.1] (low-medium), [202.4, 1181.3] (medium), [0.0, 821.9] (medium-high), [0.0, 498.3] (high) × 10^6 m^3. Besides, the characteristics of water shortage amount would vary in different seasons. Under case 1 at the low level, it would be [696.9, 1642.7] ($t = 2$, 2021-2025), [664.1, 1719.0] ($t = 5$, 2036-2040), [936.3, 1902.2] ($t = 6$, 2041-2045), [809.3, 1696.0] ($t = 10$, 2061-2065), and [938.9, 1925.4] ($t = 11$, 2066-2070) × 10^6 m^3. Generally, the water shortage would occur in future, which is affected not only by changes of climate factors, but also is connected with water conservancy project interception.

IOP Conf. Series: Earth and Environmental Science **356** (2019) 012020 doi:10.1088/1755-1315/356/1/012020

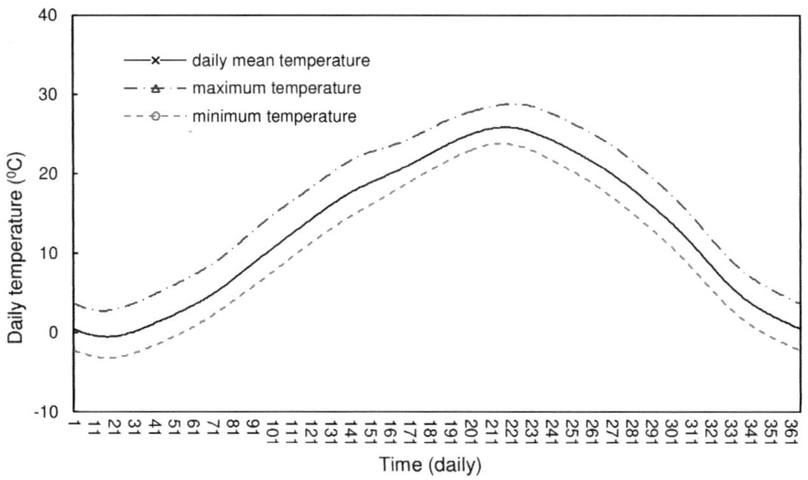

Figure 1. Daily air temperature during 1981-2010

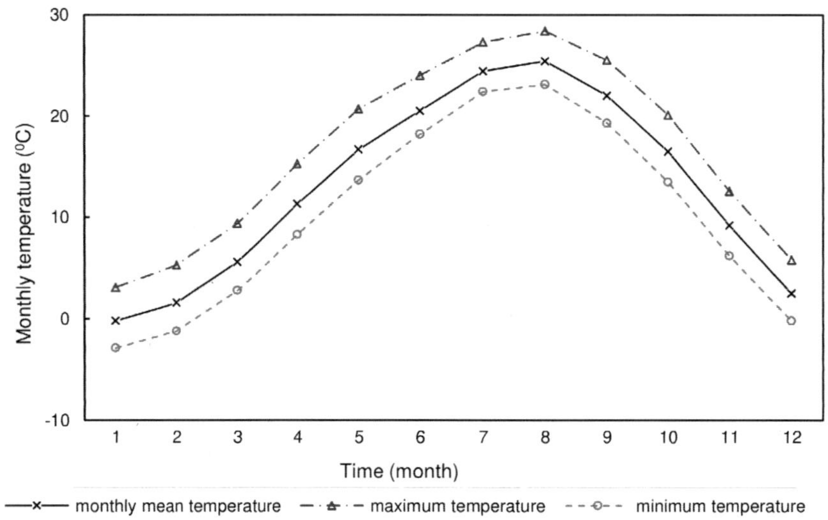

Figure 2. Monthly air temperature during 1981-2010

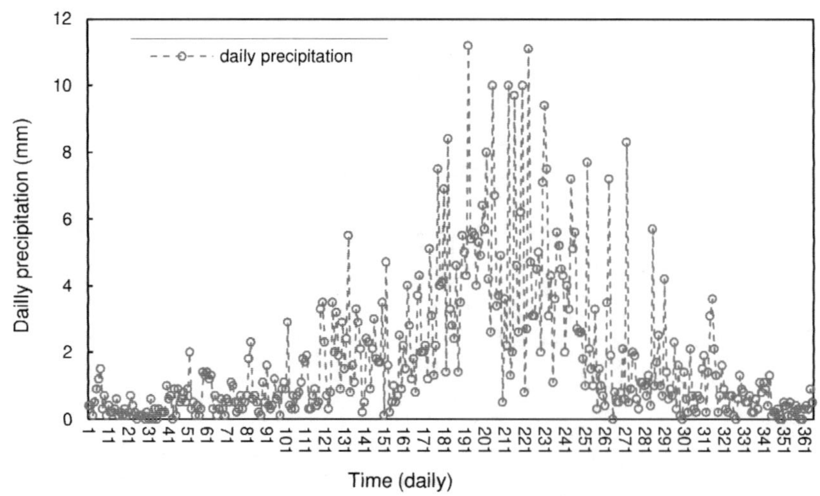

Figure 3. Daily precipitation during 1981-2010

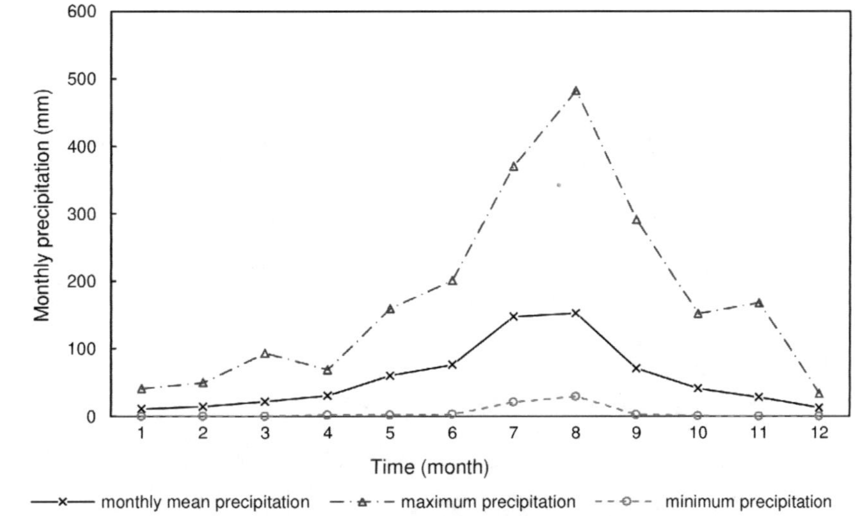

Figure 4. Monthly precipitation during 1981-2010

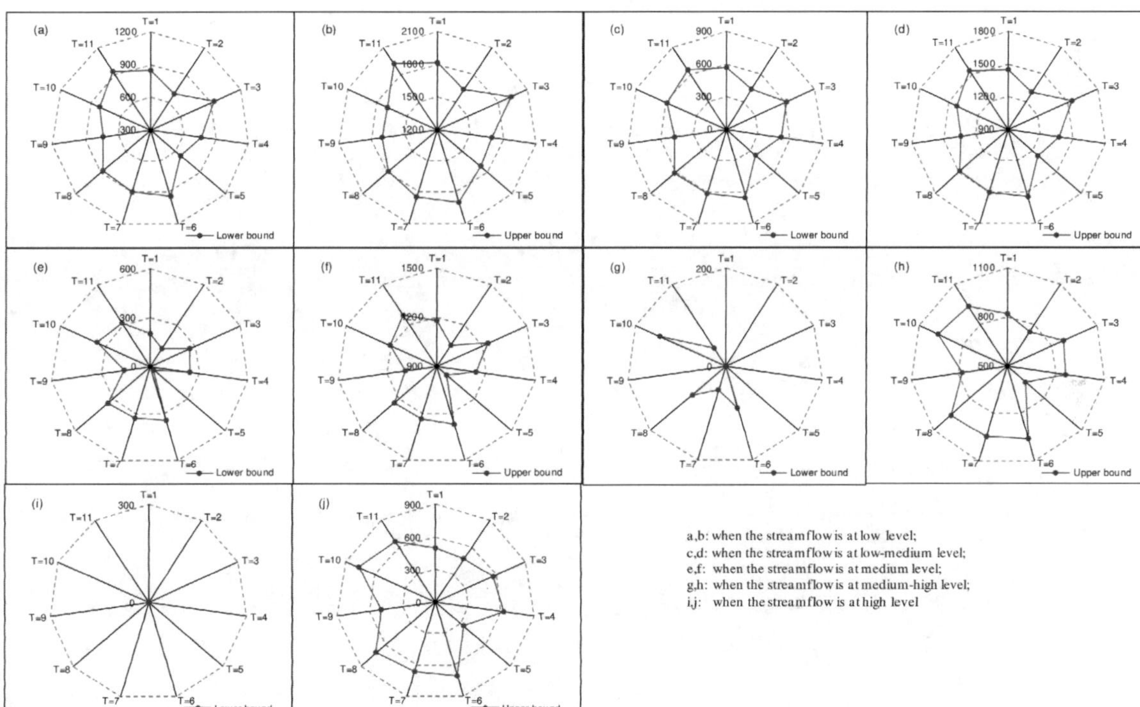

Figure 5. The water shortage amount (10^6 m^3) under case 1 (q = 0.01) when the streamflow is at low, low-medium, medium, medium-high, and high levels

4. References

[1] IPCC 2007. Freshwater resources and their management. Climate change 2007: impacts, adaptation and vulnerability. Contribution of working group II to the fourth assessment report of the intergovernmental panel on climate change. Cambridge University Press, Cambridge.

[2] Semenov MA, Barrow EM, 1997. Use of a stochastic weather generator in the development of climate change scenarios. Climate Change, 35: 397-414.

[3] Arnell NW, 1999. Climate change and global water resources. Global Environmental Change, 9 (S1), 31-49.

[4] Zhuang XW, Li YP, Huang GH, Zeng XT, 2015. An inexact joint-probabilistic programming method for risk assessment in water resources allocation. Stochastic Environmental Research and Risk Assessment, 29: 1287-1301.
[5] Huang GH, Huang YF, Wang GQ, Xiao HN, 2006. Development of a forecasting system for supporting remediation design and process control based on NAPL biodegradation simulation and stepwise-cluster analysis. Water Resources Research, 42: W06413.
[6] Nash JE, Sutcliffe JV, 1970. River flow forecasting through conceptual models, part I-a discussion of principles. Journal of Hydrology, 10: 282-290.
[7] Teutschbein C, Wetterhall F, Seibert J, 2011. Evaluation of different downscaling techniques for hydrological climate-change impact studies at the catchment scale. Climate Dynamics, 37: 2087-2105.
[8] Sperber KR, Annamalai H, Kang IS, Kitoh A, Moise A, Turner A, Wang B, Zhou T, 2013. The Asian summer monsoon: an intercomparison of CMIP5 vs. CMIP3 simulations of the late 20th century. Climate Dynamics, 41: 2711-2744.
[9] Regonda SK, Rajagopalan B, Clark M, Pitlick J, 2005. Seasonal cycle shifts in hydroclimatology over the western United States. Journal of Climate, 18(2): 372-384.
[10] Labat D, Godderis Y, Probst JL, Guyot JL, 2004. Evidence for global runoff increase related to climate warming. Advances in Water Resources, 27(6): 631-642.

Acknowledgement
This research was supported by the Science and Technology Foundation of Shandong Province (ZR2018BD005), the National Natural Science Foundation for Distinguished Young Scholar (51225904).

AUTHOR INDEX

Institute of Physics
Dirac House, Temple Back
Bristol BS1 6BE UK

ISSN: 1755-1307
ISBN 978-1-7138-0152-8

CURRAN ASSOCIATES INC.
proceedings.com